The Changing Face of
LEARNING TECHNOLOGY

Edited by

David Squires, Gráinne Conole and Gabriel Jacobs

UNIVERSITY OF WALES PRESS · CARDIFF 2000

© Association for Learning Technology, 2000

British Library Cataloguing-in-Publication Data.
A catalogue record for this book is available from the British Library.

ISBN 0–7083–1681–6

Typeset at University of Wales Press
Printed in Great Britain by Dinefwr Press, Llandybïe

Contents

Learning technology in a networked infrastructure

The future

Introduction: the changing face of learning technology

David Squires, Gráinne Conole and Gabriel Jacobs

Introduction

Technology has changed during the seven-year lifetime of *ALT-J*. In 1993 few would have predicted the pervasive influence of the Web, while JISC's vision of a Distributed National Electronic Resource (DNER) would have been regarded as little more than a flight of fancy. But has technology led to the development and adoption of genuinely innovative approaches to learning and teaching? Do teachers now teach differently and foster new relationships with students as a result of technological developments? Has learning technology overcome the criticism of failing to deliver on its promises? Do institutions function in a different way by using technology to respond to the needs of a more diverse student population and the calls for lifelong learning? Has learning technology made it possible to promote contemporary theories of learning? We aim in this volume to provide a historical perspective on learning technology which will allow readers to reflect on these and similar questions.

The papers have been chosen as being representative of the changes in learning technology over the last seven years (although it goes without saying that the choice has been a challenging one). The selection has been guided by four themes which emerged from our review of papers published in *ALT-J*: design and evaluation of technology-mediated learning environments, institutional change, learning technology in a networked infra-structure, and reflections on future possibilities. We have attempted to select papers that illustrate developments in each of these areas, and to cover the full range of the *ALT-J* publication period. In keeping with this historical perspective, and with the exception of the most recently published papers, we asked authors to write short updates, highlighting any development of the ideas they originally presented.

Design and evaluation of learning technology

Grabinger and Dunlap set the tone for our historical review (see also Grabinger *et al*, 1997). In their description of Rich Environments for Active Learning, they elucidate a philosophy for the design of technology-mediated learning environments, a philosophy predicated on an acceptance of a constructivist view of knowledge. Fowler and Mayes provide a recent analysis of the relationship between design and contemporary theories of learning, stressing the social context of learning to introduce the concept of a 'learning relationship' which 'attempts to bridge the psychological and anthropological views of situated learning'.

Informed evaluation of the effects of the use of learning technology is clearly crucial, and reliable research into evaluation methodology therefore essential. The point is forcefully underlined by Mitchell in his critique of current evaluation practice. In his view, 'much published research about education and the impact of technology is pseudo-scientific; it draws unwarranted conclusions based on conceptual blunders, inadequate design, so-called measuring instruments that do not measure, and/or use of inappropriate statistical tests'. If he is right, there is an overwhelming case for the learning-technology community to put its house in order.

Gunn partly takes up the concerns expressed by Mitchell in her proposal for a conceptual approach to evaluation. She calls for the development of evaluation techniques which resonate with practice in the use of learning technology, in other words qualitative approaches which tease out the rich context-specific aspects of learning-technology usage. Other contributors to *AltJ* have echoed this approach. For example, Draper (1997) has advocated 'niche-based' evaluation which is highly tuned to the specific contexts in which learning technology is used. Such papers draw attention to the decisive importance of developing appropriate evaluation methodologies, an importance which can only increase with the growing diversity of learning technology.

Institutional change

If learning technology is to realize its potential, it must be integrated into the daily practice of higher education. This implies that institutions will need to change in order to accommodate the new possibilities, an issue that has been a recurrent theme in *ALT-J* (see, for example, Harrison, 1994; Longstaffe *et al*, 1996; Conole and Oliver, 1998). We have chosen four papers to illustrate the focal significance of institutional change.

The paper by McCartan *et al* illustrates initial reactions to the prospect of institutional change. The question being asked was: 'How can learning technology fit into existing structures?' The question increasingly being asked now is: 'How can institutional structures be modified to accommodate the possibilities afforded by technology?', an approach evident in the title of the paper by Brown: 'Reinventing the university'. Recent changes in the student population, with students coming from more diverse backgrounds, together with a greater emphasis on flexible forms of delivery, have highlighted issues of how institutional structures can be remoulded to exploit the new technologies. Brown points out that there are alternatives to the radical notion of 'virtual universities'. And indeed, a more realistic – and effective – change may be the emergence of hybrid universities that feature both traditional and technology-mediated provision. The recent UK government's decision to broker the formation of a national eUniversity, based on an amalgamation of existing conventional higher-education institutions, bears witness to this policy. Hart and colleagues provide an illustration of how learning technology can be integrated into curricula without simply jettisoning existing practice. Their institution adopted a flexible delivery policy with the main objective of developing a more student-centred approach to teaching and learning, given that its students are predominately part-time, mature-age, already in employment and from diverse backgrounds.

Of course, learning technology will never be fully integrated into higher education unless academics come to believe in it. Continuing professional development has thus always been a concern, hence a number of papers published in this area in *ALT-J* (see, for example, Littlejohn and Stefani, 1999). McNaught and Kennedy go one step further than most, claiming an indisputable pivotal role for continuing professional development. They describe how staff development forms the rationale for a 'bottom-up' approach to realizing the objective of a top-down institutional learning technology strategy.

Learning technology in a networked infrastructure

The claims for using networks to support learning and teaching are many and various, and *ALT-J* has featured several papers on this topic. For example, Mozzon-McPherson (1996) describes the use of email to assist in teaching Italian, Brailsford *et al* (1997) and Dineen *et al* (1999) treat the design and use of software to reconfigure past electronic discussions as teaching material, and Nicholson (1998) discusses the provision of flexible learning using the Web and computer conferencing.

The three papers we have selected as representative of this category give an overview of the possibilities afforded by networked learning environments. Naida and colleagues discuss some comparatively early attempts to use asynchronous computer-mediated communication to deliver instructional content. The points they make are to be compared with the results of a recent study of the use of computer conferencing described in the paper by Jones. And Minasian-Batmanian and colleagues, in their description of an online problem-based learning approach to the creation of a student-centred learning environment, explain how a state-of-the-art networked infrastructure can play a major role in offering lifelong learning opportunities.

The future

Some would argue that the rapid rate of technological change makes any speculation about the future of learning technology a fruitless exercise. We nevertheless believe that such speculation can inform research and development agendas, and have therefore included papers concerned with potential future developments. The review by Gardner and Ward of BT's Real-time Interactive Social Environment (RISE) illustrates the possible shape of things to come. RISE is a coherent online environment offering students, teachers and support staff a virtual classroom, public and private information spaces, a shared diary, access to a modular set of coursework, email and text-based discussion groups, and personalized home pages. Squires even looks forward in his paper to the notion of the Peripatetic Electronic Teacher existing solely or partly as a telepresence in distributed multimedia environments.

Learning technology has changed significantly in the seven years since the first issue of *ALT-J* appeared and, on the whole, for the better. There can be little doubt that it is impinging on, even becoming an integral part of, the provision of higher education, enhancing practice and providing the means for innovation. One of the most important emergent themes appears to be the way in which learning technology has acted as a catalyst within institutions, encouraging both managers and lecturers to reflect on all aspects of the learning and teaching provision. It will be interesting to see if this effect continues in years to come. Let us hope that readers who casually pick up this volume seven years from now will not deride the efforts of the 1990s pioneers it features, nor find the prognostications too laughable.

References

Brailsford, T. J., Davies, P. M. C., Scarborough, S. C. and Trewhella, W. J. (1997), 'Knowledge tree: putting discourse into computer-based learning', *ALT-J,* 5 (1), 19–26.

Conole, G. and Oliver, M. (1998), 'A pedagogical framework for embedding C&IT into the curriculum', *ALT-J,* 6 (2), 4–16.

Dineen, F., Mayes, J. T. and Lee, J. (1999), 'Vicarious learning through capturing task-based discussions', *ALT-J,* 7 (3), 33–43.

Draper, S. W. (1997), 'Prospects for the summative evaluation of CAL in higher education', *ALT-J,* 5 (1), 33–9.

Grabinger, S., Dunlap, J. C. and Duffield, J. A. (1997), 'Rich environments for active learning in action: problem-based learning', *ALT-J,* 5 (2), 5–17.

Harrison, C. (1994), 'Role of learning technology in planning change in curriculum delivery and design', *ALT-J,* 2 (1), 30–7.

Littlejohn, H. and Stefani, L. A. J. (1999), 'Effective use of communication and information technology: bridging the skills gap', *ALT-J,* 7 (2), 66–77.

Longstaffe, J. A., Williams, P. J., Whittlestone, K. D., Hammond, P. M. and Edwards, J. (1996), 'Establishing a support service for educational technology within a university', *ALT-J,* 4 (1), 85–92.

Mozzon-McPherson, M. (1996), 'Italian via email: from an online project of learning and teaching towards the development of a multi-cultural discourse community', *ALT-J,* 4 (1), 40–50.

Nicholson, B. (1998), 'A case study of campus-based flexible learning using the World Wide Web and computer conferencing', *ALT-J,* 6 (3), 38–46.

Editors

David Squires is a Professor of Educational Computing, School of Education, King's College London. His research and development interests are the design and evaluation of educational software, teaching in networked environments and the use of ICT in academic research. He is editor of *ALT-J* and a member of ALT's Information Executive. [*david.squires@kcl.ac.uk*].

Gráinne Conole is Institute Director of the Institute for Learning and Research Technology, University of Bristol. Her research and development interests are the evaluation and implementation of ICT in higher education contexts. She is deputy editor of *ALT-J* and a member of ALT's Information Executive.
[*g. conole@bristol. ac. uk*]

Gabriel Jacobs is Professor of Business Management in the European Business Management School, University of Wales Swansea. He is a founder member of ALT (and indeed, with Graham Chesters and Jonathan Darby, first conceived the idea of the association). He is Chair of ALT's Information Executive, and from the beginning has been a member of the Central Executive. He was the Editor of *ALT-J* from 1993 until 1998, and remains as Executive Editor. His research interests focus on business informatics, computer-assisted learning (and, in his spare time, modern French literature and culture).
[*g. c. jacobs@ swansea. ac. uk*]

Design and evaluation of learning technology

Rich environments for active learning: a definition

R. Scott Grabinger and Joanna C. Dunlap
University of Colorado at Denver

Initially published in 1995

Rich Environments for Active Learning, or REALs, are comprehensive instructional systems that evolve from and are consistent with constructivist philosophies and theories. To embody a constructivist view of learning, REALs:

- *promote study and investigation within authentic contexts;*
- *encourage the growth of student responsibility, initiative, decision-making, and intentional learning;*
- *cultivate collaboration among students and teachers;*
- *utilize dynamic, interdisciplinary, generative learning activities that promote higher-order thinking processes to help students develop rich and complex knowledge structures; and*
- *assess student progress in content and learning-to-learn within authentic contexts using realistic tasks and performances.*

REALs provide learning activities that engage students in a continuous collaborative process of building and reshaping understanding as a natural consequence of their experiences and interactions within learning environments that authentically reflect the world around them. In this way, REALs are a response to educational practices that promote the development of inert knowledge, such as conventional teacher-to-student knowledge-transfer activities.

In this article, we describe and organize the shared elements of REALs, including the theoretical foundations and instructional strategies to provide a common ground for discussion. We compare existing assumptions underlying education with new assumptions that promote problem-solving and higher-level thinking. Next, we examine the theoretical foundation that supports these new assumptions. Finally, we describe how REALs promote these new assumptions within a constructivist framework, defining each REAL attribute and providing supporting examples of REAL strategies in action.

Introduction

In today's complex world, simply knowing how to use tools and knowledge in a single domain is not sufficient to remain competitive as either individuals or companies. People must also learn to apply tools and knowledge in new domains and different situations. Industry specialists report that people at every organizational level must be creative and flexible problem solvers (Lynton, 1989). This requires the ability to apply experience and knowledge to address novel problems. Consequently, learning to think critically, to analyse and synthesize information to solve technical, social, economic, political, and scientific problems, and to work productively in groups are crucial skills for successful and fulfilling participation in our modern, competitive society.

The purpose of this article is to describe and organize the shared elements of Rich Environments for Active Learning, or REALs, including the theoretical foundations and instructional strategies to provide a common ground for discussion. REALs are based on constructivist values and theories including 'collaboration, personal autonomy, generativity, reflectivity, active engagement, personal relevance, and pluralism' (Lebow, 1993, p. 5). REALs provide learning activities that, instead of transferring knowledge to students, engage students in a continuous collaborative process of building and reshaping understanding as a natural consequence of their experiences and authentic interactions with the world (Goodman, 1984; Forman and Pufall, 1988; Fosnot, 1989). Advocating a holistic approach to education, REALs reflect the assumption that the process of knowledge acquisition is 'firmly embedded in the social and emotional context in which learning takes place' (Lebow, 1993, p. 6).

Need for educational change

Changing society
Education is receiving increasing pressure from changing global economic circumstances and complex societal needs. Yet, according to Lynton (1989, p. 23), 'at this time [...] education is far from fully contributing to the economic well-being of this country [United States]'. Public and private institutions are demanding employees who can think critically and solve a range of problems, who can move easily from task to task, and who can work efficiently and effectively in team situations; yet they claim that those people are difficult to find.

The US education system, to its credit, is neither deaf to this plea nor ignorant of the need to change how and what instruction is delivered in the classroom. Calls for restructuring the way students learn come from a variety of institutions including the American Association for the Advancement of Science (1989) and the National Council of Teachers for Mathematics (1989). Educators agree that students must learn to solve problems and think independently (Feuerstein, 1979; Mann, 1979; Segal *et al*, 1985; Linn, 1986; Resnick and Klopfer, 1989; Bransford *et al*, 1990). The challenge for educators is to utilize strategies that teach content in ways that also help to develop thinking, problem-solving, metacognitive, and life-long learning skills (Bransford *et al*, 1990; Savery and Duffy, 1994).

Weaknesses within the current system
There is considerable evidence that today's students are not particularly strong in the areas of thinking and reasoning (Resnick, 1987; Nickerson, 1988; Bransford *et al*, 1991). Bransford *et al* (1990, pp. 115–16) state that the 'basic problem is that traditional instruction often fails to produce the kinds of transfer to new problem-solving situations that most educators would like to see'. Conventional instruction often utilizes simplified,

decontextualized examples and problems, leading to an inadequate understanding of and ability to apply the knowledge acquired. Students often are not exposed to examples and problems that make knowledge relevant to them (Collins *et al*, 1991). Instead, students are asked to solve problems that cause them to wonder: 'Why do I need to know this?'. Because the information presented to students has no relevance or meaning for them, they tend to treat new information as facts to be memorized and recited rather than as tools to solve problems relevant to their own needs (Bransford *et al*, 1990). This, unfortunately, leads to inert knowledge – knowledge that cannot be applied to real problems and situations.

Inert knowledge

Research shows that knowledge learned but not explicitly related to relevant problem-solving remains inert (Whitehead, 1929; Perfetto *et al*, 1983, CTGV, 1993c). Whitehead first coined the phrase *inert knowledge* in 1929 to refer to knowledge acquired in abstract circumstances without direct relevance to the learner's needs. Inert knowledge is not readily available for application or transfer to novel situations (for a review of transfer research, refer to Clark and Voogel, 1985; Butterfield and Nelson, 1989). It seems that our educational system is focused on producing inert knowledge. The Cognition and Technology Group at Vanderbilt (CTGV) (CTGV, 1993c) specifies the following flaws in our conventional approaches to schooling and teaching that lead to inert knowledge:

- There is a constant battle of breadth versus depth – and breadth usually wins. We (educators) tend to fill our students with facts, and leave no time for dealing with topics in depth. We expect our students to remember dates, formulae, algorithms, quotations, and whole poems, yet show little practical use for that knowledge despite the fact that we know our students have difficulty transferring the knowledge. Robertson (1990) states that:

    Students who rely on memorized algorithms for solving problems typically do not perform as well on transfer problems as do students who rely on an understanding of the underlying concepts. (p. 253)

- We consistently rely on decontextualized instructional strategies. In our desire to cover as much material as possible, we focus our instructional activities on abstract basic skills, concepts, and technical definitions. We believe that decontextualized skills have broad applicability and are unaffected by the activities or environments in which they are acquired and used (Brown *et al*, 1989). However, when we do this, students do not learn when to apply those skills or within what kinds of contexts they work. We do this despite a large body of evidence that indicates that abstracted skills are seldom transferred from one domain to another (Clark and Voogel, 1985; Butterfield and Nelson, 1989). We fail to realize that abstracted skills have no contextual cues to relate to new situations.

- When we do provide practice for our students, we give them arbitrary, uninteresting, unrealistic problems to solve. The example of story problems in mathematics is over-used. We can also find examples of unrealistic and over-simplified problems in the sciences, languages, and social studies. Again, we do this in the mistaken belief that we must emphasize decontextualized skills that are applicable everywhere. Yet these unrealistic problems have no meaning to the students who then fail to find any contextual cues to relate to problems they may encounter in their lives.

- We treat students passively for 12 to 16 years, rarely giving them the opportunity to take responsibility for their own learning, to explore ideas of their own choosing, to collaborate with one another or with teachers, or to make valuable contributions to the learning of others. They do not learn to take charge of their own learning, nor do they learn the skills necessary to become life-long learners and daily problem-solvers.

To the preceding four items cited in the CTGV article, we add a couple more to the list of conventional educational practices that foster the production of inert knowledge:

- Students are not evaluated in authentic ways. After teaching in decontextualized ways, we test in the same ways. We do not look at actual performance but use paper-pencil tests to measure the quantity of knowledge learned. We fail to examine the quality of their thinking or problem-solving.

- Finally, our current school practices often have negative effects on the morale and motivation of students. Perelman (1992) states:

  > Students are forced to compete to achieve as much as they can within the periods of time allotted for each activity. This design requires that most students fail or do less well most of the time so that a minority of them can be labelled 'excellent.' The main functional focus of the system is not 'learning,' it is 'screening out'. (p. 72)

This has a tremendous effect on society. Drop-out rates are higher than our society can afford. We have created an evaluation, testing, and grading sub-structure that helps perpetuate the view that education is a game that has few winners and many losers. This game teaches our students to focus on tests and grades rather than on problem-solving, applying knowledge, and working collaboratively in a risk-free environment. The best students learn early on that they succeed best by working by themselves as quickly as possible. They learn to beat the tests and win the game.

Erroneous assumptions
We begin to change these conventional practices by calling into question some of our basic assumptions. Berryman (1991) states that the educational practices described above stem from five erroneous assumptions about learning that have governed education since the beginning of the industrial age. He holds that we often assume incorrectly that:

- people easily transfer learning from one situation to another if they have learned the fundamental skills and concepts [decontextualized];

- learners are 'receivers' of knowledge in verbal forms from books, experts, and teachers;

- learning is entirely behaviouristic, involving the strengthening of bonds between stimuli and correct responses;

- learners are blank slates ready to be written upon and filled with knowledge; and

- skills and knowledge are best acquired independent of realistic contexts.

To begin to address the issues of transfer and instructional methods to meet employer and societal needs, reasoning and problem-solving skill development must be a part of an interdisciplinary program of study in education (Lynton and Elman, 1987) – a program or environment that places students in situations where they can practise solving problems in a meaningful and constructive manner.

We need to look at other ways

One view of an alternative framework comes from researchers who are beginning to emphasize the importance of anchoring or situating instruction in meaningful problem-solving environments (CTGV, 1993c). The Cognition and Technology Group at Vanderbilt is a leader in the development of alternative frameworks of instruction and schooling. The group posits two significant changes.

First, we as educators must establish *new goals for learning*. We must move from emphasizing decontextualized reading and computational skills to developing independent thinkers and learners who engage in life-long learning. This does not mean that we abandon the important skills of reading and computation; it means instead that we should be teaching reading and computation within more situated contexts that demonstrate the value of those skills.

New assumptions about learning

Second, in contrast to our long operative conventional assumptions (see above), we must base our teaching on *new assumptions about the nature of thinking, learning, and instruction*. We must accept that

> [. . .] the mere accumulation of factual or declarative knowledge is not sufficient to support problem solving. In addition to factual or declarative knowledge, students must learn why, when, and how various skills and concepts are relevant. (CTGV, 1993c, p. 79)

Effective problem-solving and thinking are not based solely on motivation and knowledge of thinking strategies, but also on well-organized and indexed content knowledge. Learners must have rich knowledge structures with many contextual links to help them address and solve complex problems. This means that instead of trying to abstract a set of decontextualized general skills, we must make our teaching as contextualized as possible to provide as many possible links with other domains as possible.

Therefore, we propose the following 'new' assumptions about learning and teaching:

- People transfer learning from one situation to another with difficulty. Learning is more likely to be transferred from complex and rich learning situations. Rich learning activities help students think deeply about content in relevant and realistic contexts (CTGV, 1993c).

- Learners are 'constructors' of knowledge in a variety of forms. They take an active role in forming new understandings and are not just passive receptors.

- Learning is a collaborative process. Students learn not solely from experts and teachers, but also from each other. They test ideas with each other and help each other build elaborate and refined knowledge structures.

- Learning is cognitive, and involves the processing of information and the constant creation and evolution of knowledge structures. We must focus on and make visible thinking and reasoning processes. We are not suggesting abandoning the teaching of content to teach only thinking and reasoning, because 'knowledge of concepts, theories, and principles [. . .] empowers people to think effectively' (Bransford *et al*, 1990, p. 115).

- Learners bring their own needs and experiences to a learning situation and are ready to act according to those needs. We must incorporate those needs and experiences into learning activities to help students take ownership and responsibility for their own learning.

- Skills and knowledge are best acquired within realistic contexts. Morris *et al* (1979) call this 'transfer appropriate processing.' Transfer appropriate processing means that students must have the opportunity to practise and learn the outcomes that are expected of them under realistic or authentic conditions.

- Assessment of students must take more realistic and holistic forms, utilizing projects and portfolios and de-emphasizing standardized testing. Educators are increasingly aware that conventional achievement and intelligence tests do not measure the ability of people to perform in everyday settings and adapt to new situations (CTGV, 1993c).

A discussion of the foundations for these assumptions and examples of their implementation makes up the rest of this article.

Rich Environments for Active Learning

Definition of REALs

We must implement a number of strategies in order successfully to adopt the new assumptions about thinking, learning, instruction, and achievement described above. The adoption of these strategies creates learning environments that we call Rich Environments for Active Learning (REALs). REALs are comprehensive instructional systems that (Dunlap and Grabinger, 1992, 1993; Grabinger and Dunlap, 1994a, 1994b):

- evolve from and are consistent with constructivist philosophies and theories;

- promote study and investigation within authentic (i.e. realistic, meaningful, relevant, complex, and information-rich) contexts;

- encourage the growth of student responsibility, initiative, decision-making, and intentional learning;

- cultivate an atmosphere of knowledge-building learning communities that utilize collaborative learning among students and teachers (Collins, 1995);

- utilize dynamic, interdisciplinary, generative learning activities that promote high-level thinking processes (i.e. analysis, synthesis, problem-solving, experimentation, creativity, and examination of topics from multiple perspectives) to help students integrate new knowledge with old knowledge and thereby create rich and complex knowledge structures; and,

- assess student progress in content and learning-to-learn through realistic tasks and performances.

It is important to note that two of the overall defining characteristics of learning environments are integration and comprehensiveness (Hannafin, 1992). Hannafin describes *integration* as a process of linking new knowledge to old, and modifying and enriching existing knowledge. Integration enhances the depth of learning to increase the number of access points to that information. As we described earlier, learning environments must go beyond general, abstracted learning to include specific learning. Goldman *et al* state that:

> These environments are designed to invite the kinds of thinking that help students develop *general* skills and attitudes that contribute to effective problem solving, plus acquire *specific* concepts and principles that allow them to think effectively about particular domains. (p. 1)

Comprehensiveness, the second defining characteristic, carries this notion further. It refers to the importance of placing learning in broad, realistic contexts rather than in decontextualized and compartmentalized contexts. REAL learning strategies, then, guide and mediate an individual's learning, and support the learner's decision-making (Hannafin, 1992). Themes are used to help organize learning around interdisciplinary contexts that focus on problem-solving or projects that link concepts and knowledge to focused activities within the environment (Hannafin, 1992).

What a REAL is not

Because the phrase *learning environment* is broadly and carelessly used in educational literature to describe everything from schools to classrooms to computer microworlds to learning activities to air conditioning and furniture, we shall try to clarify what a REAL is not before examining the attributes in more detail. We attempt to make the case that a REAL is a more accurate description of what people generally mean when they use the term *learning environment*.

First, a REAL is not a delivery technology like video, CD-ROM, or audio tapes. Clark (1994) defines *delivery technologies* as those that draw on resources and media to deliver instruction, and affect the cost and access of instruction. Media technologies can be integral components of REALs and, in fact, usually are. However, a REAL is not limited to any specific media, but instead is an assortment of methods and ideas that help create an environment that promotes and encourages active learning. Clark's point is important from a research standpoint because instructional methods are often confounded with media in research, and he argues strongly and convincingly that it is instructional methods, not media, that influence learning. He contends that any necessary teaching method can usually be designed in more than one medium. Although there are varying degrees of acceptance and disagreement with Clark's point of view (for example, Jonassen, 1994b; Kozma, 1994), a REAL is a set of instructional methods designed on the assumptions that media are tools for students and teachers to use, and that the learning that occurs within the environment is founded on the activities and processes that encourage thinking and reasoning, not the media that deliver information.

Second, do not confuse REALs with computer-based microworlds or learner-support environments (Allinson and Hammond, 1990). Computer-based microworlds are computer programs that are designed to apply constructivist theories. Examples include case-based applications, simulations, intentional learning environments, and some hypermedia resources. Developers of microworlds often refer to their programs as learning environments because they often attempt to simulate, on a smaller simplified scale, realistic environments. However, we contend that this limits the concept of learning environment. Learning environments, and especially REALs, are much more comprehensive and holistic than individual computer applications. Although some computer-based applications use constructivist ideas quite admirably – see especially the *Strategic Teaching Framework* (Duffy, 1992; Duffy, in press), and the *Transfusion Medicine Modules* (Ambruso and The Transfusion Medicine Group, 1994) – they are not learning environments in the sense that REALs are. To create REALs, teachers must involve their students, parents, administrators, and colleagues in planning and implementing strategies that encourage student responsibility, active knowledge construction, and generative learning activities on a large scale and in a variety of methods and forms. Microworlds may play a role in a REAL through the delivery of information, practice, finding and presenting information, stimulation of high-level thought processes, promotion of collaboration, or exploration; but REALs involve many more activities and demand much more flexibility than can probably ever be contained in a single computer program. A REAL is, then, a learning community that 'includes the content taught, the pedagogical methods employed, the sequencing of learning activities, and the sociology of learning' (Collins *et al*, 1991, p. 6).

Finally, a REAL is not the physical attributes of the classroom. It is not about air conditioning, desks, lighting, or ergonomics. While these environmental factors are critical in creating a smooth-operating learning environment, they are not included in our definition of REALs (for information about this topic, see Gayeski, 1995).

Foundation of REALs

REALs (i.e. constructivist learning environments, information-rich learning environments, interactive-learning environments, or knowledge-building learning communities) are not new to education. We can go back to Socrates and see that he used problems and questions to guide students to analyse and think about their environments (Coltrane, 1993). Rousseau prescribed using direct experience (Farnham-Diggory, 1992). In the early 1900s, John Dewey (1910) proposed student-directed reforms and experiential learning. Bruner (1961) advocated discovery or inquiry learning around realistic problems. The notion that students should learn through practice, application, and apprenticeship has been with us for centuries. It was not until the industrial age, when we needed places to store children until old enough to work on assembly lines, that we began trying to mass produce replicable results. Yet the last five to ten years have seen renewed emphasis on reforming schools and teaching practices to replace our educational production lines with classrooms that teach people to think and solve problems. This current effort at renewal revolves around a set of ideas and theories referred to as constructivism.

Characteristics of constructivism

The class of theories that guides the development of REALs is called constructivist theories (Clement, 1982; Bransford and Vye, 1989; Minstrell, 1989; Resnick and Klopfer, 1989; Schoenfeld, 1989; Bednar *et al*, 1991; Duffy and Bednar, 1991; Perkins, 1991; Scardamalia and Bereiter, 1991; Spiro *et al*, 1991). Fundamentally, constructivism asserts that we learn through a continual process of building, interpreting, and modifying our own representations of reality based upon our experiences with reality (Jonassen, 1994c).

Wheatley (1992) elaborates on the importance of this idea not just in learning but also in the innovation and improvement of society:

> Learning (innovation) is fostered by information gathered from new connections; from insights gained by journeys into other disciplines or places; from active, collegial networks and fluid, open boundaries. Learning (innovation) arises from ongoing circles of exchange where information is not just accumulated or stored, but created. Knowledge is generated anew from connections that weren't there before. When this information self-organizes, learning (innovation) occurs, the progeny of information-rich, ambiguous environments. (p. 113)

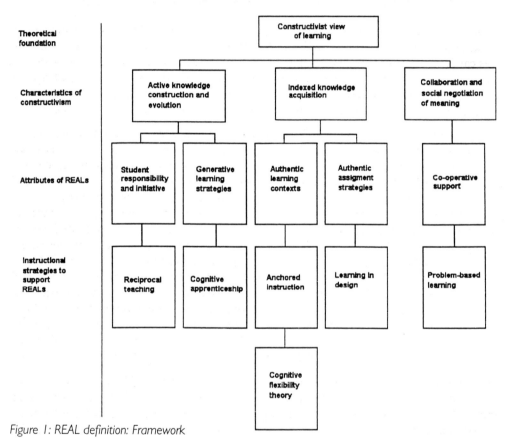

Figure 1: REAL definition: Framework

To develop learning environments that encourage the type of learning described by Jonassen, Wheatley, and others, REALs reflect the theories and philosophies of constructivism. In fact, the three main characteristics of the constructivist view of learning govern our design of REALs (see Figure 1). The first characteristic of constructivism is the notion that knowledge is not a product to be accumulated but an active and evolving process in which the learner attempts to make sense out of the world (Gurney, 1989). Brown *et al* (1989) illustrate this idea:

A concept, for example, will continually evolve with each new occasion of use, because new situations, negotiations, and activities inevitably recast it in a new, more densely textured form. So a concept, like the meaning of a word, is always under construction. (p. 33)

The second characteristic is the notion that people conditionalize their knowledge in personal ways (Gurney, 1989). That is, they acquire knowledge in forms that enable them to use that knowledge later. Bransford *et al* (1990) state that:

[. . .] there are large differences between knowing something and spontaneously thinking to do it or use it when one is engaged in an actual problem-solving situation. (p. 122)

Knowledge, then, is 'indexed' to the contexts in which we encounter it. We are unlikely to use knowledge that is decontextualized because we can see no relevance – no contextual cues cause us to remember that knowledge because it was taught without cues. A person who indexes and conditionalizes knowledge knows when to apply that knowledge. A person who learns in a decontextualized way often is not aware that he or she has the applicable knowledge to solve a problem. In other words, knowledge acquired is inert.

Students must acquire concepts and theories in ways that help them use the information later on, and appreciate the value of that information. Brown *et al* (1989, p. 36) describe this process as *indexicalizing knowledge*. They mean that rich involvement in realistic and relevant problem-solving allows learners to develop many broad and deep indexicalized representations that enable them to apply knowledge more spontaneously to new situations because they can compare a known and relevant situation with a new situation. The more links there are across related knowledge structures, the more likely students are to apply that knowledge. The more closely the learning context resembles the actual context, the better people will perform. Tulving and Thompson (1973) refer to this as 'encoding specificity,' which holds that successful retrieval of information is enhanced when cues relevant to later retrieval of that information are encoded along with the material learned. It is important to note that constructivists contend that these rich links cannot be developed in decontextualized learning activities; rather, learning must be placed in realistic contexts that provide cognitive conflict or puzzlement, and determine the organization and nature of what is learned (Savery and Duffy, 1994).

White (see Robertson, 1990) theorizes that there are two kinds of links that need to be developed while learning: internal and external. Internal associations are connections among the criterial attributes of a principle. Internal associations reflect the learner's understanding of the concept. External associations refer to connections between the principle and everyday experiences or context, and indicate the 'usability' of a concept. Learning to solve problems requires both kinds of links. Our schools are good at building the internal links, but poor at providing the external links.

The third major characteristic of constructivism is the importance of collaboration and social negotiation of meaning. Learning happens within a social context. Conceptual growth comes from sharing our perspectives and testing our ideas with others, modifying our internal representations in response to that process of negotiation (Bednar *et al*, 1991). Common understandings and shared meanings are developed through interaction among peers and teachers. This is the cultural aspect of knowledge.

The activities of a domain are framed by its culture. Their meaning and purpose are socially constructed through negotiations among present and past members. Activities thus cohere in a way that is, in theory if not always in practice, accessible to members who move within the social framework. These coherent, meaningful, and purposeful activities are authentic, according to the definition of the term we use here. (Brown *et al*, 1989, p. 34)

This social aspect of constructivism is important on an individual level as well as cultural level, for collaborative interactions allow us to test the viability of our understandings, theories, and conjectures (Savery and Duffy, 1994).

REALs embody the characteristics of constructivism, but how? What are the critical attributes of REALs that support a constructivist view of learning? The next section of this article discusses each of the criterial attributes of a REAL:

A. Student responsibility and initiative.

B. Generative learning activities.

C. Authentic learning contexts.

D. Authentic assessment strategies.

E. Co-operative support.

We will illustrate each attribute with examples of functioning REALs utilizing the strategies of reciprocal teaching, cognitive apprenticeship, anchored instruction, cognitive flexibility theory, learning in design, and problem-based learning (PBL).

The main attributes of REALs

Student responsibility and initiative
The first attribute of REALs is that they are student-centred. Student-centred learning environments place a major emphasis on developing intentional learning and life-long learning skills. These skills include the abilities to construct higher-order questions to guide learning, reflect on consequences and implications of actions, and monitor and modify personal cognitive activity. This attribute is critical because students cannot actively construct and evolve their knowledge structures without taking responsibility and initiative for their learning.

Intentional learning
Scardamalia and her colleagues (1989) noticed that passive or immature learners have certain characteristics that prevent them from becoming skilful problem solvers. First, immature learners tend to organize their mental activities around topics rather than goals, promoting decontextualization and the inability to see the relevance of the learning activity to their lives. Second, they tend to focus on surface features, and do not examine a topic in depth. Third, they tend to work until a task is finished. They do not take time to examine the quality of their work, nor do they take the time to consider revisions to their finished product or to the strategies utilized to complete the task. When the task is finished, they forget whatever learning took place during their work on the task. Finally, they think of learning in an additive fashion rather than transforming and enriching their existing knowledge structures.

These characteristics are, in essence, the product of the conventional kind of schooling that we described earlier. These behaviours prevent students from transferring their knowledge to new problems because they have learned content and strategies in a decontextualized context. Palincsar (1990) states that:

> To achieve transfer it is necessary to attend to the context in which instruction and practice occur; transfer is likely to occur to the extent that there are common elements between the situation in which the children are learning this tactic and the situations in which such a tactic would be useful. (p. 37)

Palincsar, Scardamalia, and Bereiter are leading proponents of the conviction that students must be taught to take more responsibility for their own learning to enhance the likelihood of transfer. They refer to this concept as intentional learning, or 'those cognitive processes that have learning as a goal rather than an accidental outcome' (Bereiter and Scardamalia, 1989, p. 363). Palincsar and Klenk (1992) state that 'intentional learning, in contrast to incidental learning, is an achievement resulting from the learner's purposeful, effortful, self-regulated, and active engagement'. To be intentional learners, students must learn to learn as well as accrue knowledge. Teaching, too, takes on a revised role, for to teach for intentional learning means to cultivate those general abilities that facilitate life-long learning (Palincsar, 1990). The main skills involved in teaching students to be more intentional are questioning, self-reflection, and metacognition, 'or the awareness and ability to monitor and control one's activity as a learner' (Brown *et al*, 1983, p. 212).

Questioning
Scardamalia and Bereiter (1991) believe that one of the first steps in developing intentional learners is to help students take more executive control over what they decide to learn through the development of higher-order questioning skills. They point out that in a typical classroom, teachers ask all the questions. Therefore, teachers are the ones engaged in the question generation process, which involves important high-level thinking skills and executive control decisions – the skills students need to develop actively to construct and evolve their knowledge structures. The students are not given an opportunity to ask questions and, therefore, do not learn to perform the analysis activities related to question generation. Research by Scardamalia and Bereiter (1991) indicates that students can learn to ask questions to guide their knowledge-building, thus assuming a 'higher level of agency' and more ownership for their learning. In a student-centred REAL, students are given more executive control over their learning to enable them to take more ownership, to find more relevance and authenticity, and to learn life-long learning skills.

Self-reflection
A second skill in intentional learning is self-reflection. '*Self-reflection* implies observing and putting an interpretation on one's own actions, for instance considering one's own intentions and motives as objects of thought' (Von Wright, 1992, p. 61). Von Wright writes that self-reflection involves the abstraction of meaning and is an interpretative process aimed at the understanding of reality. To understand the world in different ways involves modifying our conceptions of the world and our place in the world. It involves thinking about reality in alternative ways. Von Wright goes on to describe two levels of reflection. One level of reflection is the ability to think about consequences and implications of actions. A second level of reflection is the ability to think about oneself as

an intentional subject of one's own actions and to consider the consequences and efficacy of those actions. This involves the ability to look at oneself in an objective way, and to consider ways of changing future actions to improve performance. The second level of reflection also involves metacognitive learning skills.

Metacognitive skills
'Metacognitive skills refer to the steps that people take to regulate and modify the progress of their cognitive activity: to learn such skills is to acquire procedures which regulate cognitive processes' (Von Wright, 1992, p. 64). Metacognitive skills include taking conscious control of learning, planning and selecting strategies, monitoring the progress of learning, correcting errors, analysing the effectiveness of learning strategies, and changing learning behaviours and strategies when necessary (Ridley *et al*, 1992). These abilities interact with developmental maturation and domain expertise. Immature learners have not acquired the ability to regulate or modify their cognitive activity; they may have learned a single strategy, such as memorization, and default to that one strategy in all situations.

Studies show that the use of metacognitive strategies can improve learning skills, and that the independent use of these metacognitive strategies can be gradually developed in people (Brown, 1978; Biggs, 1985; Weinstein *et al*, 1988). Blakely and Spence (1990) describe several basic instructional strategies that can be incorporated into classroom activities to help students develop metacognitive behaviours:

- Students should be asked to identify consciously what they 'know' as opposed to what they 'do not know'.

- Students should keep journals or logs in which they reflect upon their learning processes, noting what works and what does not.

- Students should manage their own time and resources, including estimating time required to complete tasks and activities, organizing materials and resources, and scheduling the procedures necessary to complete an activity.

- Students must participate in guided self-evaluation using individual conferences and checklists to help them focus on the thinking process.

REAL strategy: reciprocal teaching
One of the manifestations of a REAL that emphasizes the development of intentional learning skills described above is reciprocal teaching. The context of reciprocal teaching is social, interactive, and holistic. Palincsar and Klenk (1992) used reciprocal teaching with at-risk first-grade students to develop reading skills. Palincsar and Klenk (1992) describe reciprocal teaching as:

> [. . .] an instructional procedure that takes place in a collaborative learning group and features guided practice in the flexible application of four concrete strategies to the task of text comprehension: questioning, summarizing, clarifying, and predicting. The teacher and group of students take turns leading discussions regarding the content of the text they are jointly attempting to understand. (p. 213)

These strategies are the kinds of intentional learning strategies that encourage self-regulation and self-monitoring behaviours.

The relationship of reciprocal teaching to REALs is founded on three theoretical principles, consistent with the characteristics of constructivism described previously, based on the work of Vygotsky (1978) as described by Palincsar and Klenk (1992). The first principle states that the higher cognitive processes originate from social interactions. The second principle is Vygotsky's *Zone of Proximal Development* (ZPD). Vygotsky described the ZPD as:

> [. . .] the distance between the actual developmental level as determined by independent problem-solving, and the level of potential development as determined through problem-solving under adult guidance or in collaboration with more capable peers. (p. 86)

Reciprocal teaching is designed to provide a ZPD in which students, with the help of teachers and peers, take on greater responsibility for learning activities. Finally, Vygotsky's third principle advocates that learning take place in a contextualized, holistic activity that has relevance for the learners.

How, then, does reciprocal teaching work? The process begins with a text that a class reads silently, orally, or along with the teacher depending on the skill level of the students. Following each segment, a dialogue leader (students take turns) asks questions that deal with content or 'wonderment' issues. The questions often stimulate further inquiry. The other students respond to the questions, raise their own questions, and, in cases of disagreement or confusion, re-read the text. The discussion leader is responsible for summarizing and synthesizing the reading and discussion, and clarifying the purpose of the reading. The leader also generates and solicits predictions about the upcoming text to prepare for meaningful reading of the next segment. The teacher must model the appropriate behaviour and provide scaffolding to sustain the discussion. (The preceding description is taken from Palincsar and Klenk, 1992.) The students, then, are involved in the higher-level thinking and decision-making activities that usually fall within the realm of the teacher. With the help of the teacher, students share a ZPD where they can learn the questioning, summarizing, clarifying and predicting activities so integral to metacognitive awareness.

Finally, why does reciprocal teaching work? Collins *et al* (1991) posit the following reasons for its success (and, in a broader view, for the success of REALs):

- The reciprocal teaching model engages students in activities that help them form a new conceptual model of the task of reading. They see reading as a process that involves reflection and prediction rather than just the recitation of words. They learn to make what they are reading relevant to their needs and to monitor their progress and strive for clarification.

- The teacher and student share a problem context while the teacher models expert strategies that the students learn to use independently.

- Scaffolding is crucial in the success of reciprocal teaching. 'Most importantly, it decomposes the task as necessary for the students to carry it out, thereby helping them to see how, in detail, to go about it' (Collins, *et al*, 1991, p. 11).

- Finally, students learn the self-monitoring activities and thinking processes involved in critiquing and improving their work.

Generative learning activities
The second requirement of REALs is that students engage in generative learning activities. People who learn through active involvement and use tools build an 'increasingly rich implicit understanding of the world' (Brown *et al*, 1989, p. 33). Generative learning requires that students 'engage in argumentation and reflection as they try to use and then refine their existing knowledge as they attempt to make sense of alternate points of view' (CTGV, 1993b, p. 16). Studies indicate that knowledge is more likely to be active and used when acquired in a problem-solving mode rather than in a factual-knowledge mode (Adams *et al*, 1988; Lockhart *et al*, 1988). The concept of generative learning is an extension of the constructivist characteristic of actively constructing knowledge; students cannot construct or evolve their own learning without generating something through active involvement.

Generative learning requires a shift in the traditional roles of students and instructors. Students become investigators, seekers, and problem solvers. Teachers become facilitators and guides, rather than presenters of knowledge. For example, rather than simply learning what objectives and goals are, students in a teacher-education class generate lesson plans and objectives and then manipulate and revise them to solve new teaching problems. Ideally, they test their objectives, goals, and strategies in actual practice-teaching situations. In generative learning, students *apply* the information they learn. Generative learning activities require students to take static information and generate fluid, flexible, usable knowledge. Generative learning, then, means that students are involved deeply and constantly with creating solutions to authentic problems via the development and completion of projects. A REAL strategy that relies heavily on projects is cognitive apprenticeship.

REAL strategy: cognitive apprenticeship
Cognitive apprenticeship is modelled after the traditional apprenticeship way of learning arts and crafts. It incorporates elements of traditional apprenticeship and modern schooling. In traditional apprenticeship, the processes of an activity are visible and involve learning a physical and outwardly observable activity (Collins *et al*, 1991). The expert shows an apprentice how to perform a task, then watches and coaches as the apprentice practises portions of the task, and finally turns over more and more responsibility to the apprentice until the apprentice can perform the task alone (Collins *et al*, 1991). Traditional apprenticeship deals with processes that are easily visible because they involve skills and produce visible products.

The goal of cognitive apprenticeship is to make processes that are normally invisible visible. In schooling, the process of thinking is usually invisible to both students and teachers. For example, the practices of problem-solving, reading comprehension, and computation are not visible processes (Collins *et al*, 1991). Brown *et al* (1989) point out that the term *cognitive apprenticeship* emphasizes that apprenticeship techniques can reach beyond observable physical skills to the kinds of cognitive skills associated with learning in schools. In a cognitive apprenticeship environment, the teacher attempts to make visible the thinking processes involved in performing a cognitive task. The teacher first models how to perform a cognitive task by thinking aloud. Then the teacher watches, coaches, and provides scaffolding as the students practise portions of the task. Finally, he or she turns over more and more responsibility to students, and fades coaching and scaffolding until they can perform the task alone. 'Cognitive apprenticeship supports learning in a domain by enabling students to acquire, develop, and use cognitive tools in an authentic domain activity' (Brown *et al*, 1989, p. 39).

The differences between traditional and cognitive apprenticeship (Collins *et al*, 1991) are important because they indicate where the effort must be placed on the instructional design of learning activities. First, in traditional apprenticeship the task is easily observable. In cognitive apprenticeship, the thinking processes must be deliberately brought into the open by the teacher, and the teacher must help students learn to bring their thinking into the open. Second, in traditional apprenticeship, the tasks come from work, and learning is situated in the workplace. In cognitive apprenticeship, the challenge is to situate the abstract goals of school curriculum in contexts that make sense to students. Third, in traditional apprenticeship, the skills learned are inherent in the task. In schooling, students learn skills that are supposed to transfer to different tasks. In cognitive apprenticeship, the challenge is to present a range of tasks to encourage reflection and to identify common transferable elements across tasks. The goal is to help students generalize and transfer their learning through conditionalized and indexed knowledge (related to the second characteristic of constructivism).

Cognitive apprenticeship and generative learning are closely linked because the process of making cognitive processes visible means that students must create or generate things that represent those processes. Teachers must create work and tasks that represent the process of solving a problem, writing, or computation in addition to products. To examine the development of student thinking, an English teacher may ask for questions, themes, concept maps, and outlines before students begin writing. Mathematics teachers are often notorious for telling students: 'I want to see your work, not just the answer,' so they can look for errors in the thinking process.

Generative learning is one of the simplest features of a REAL. It simply demands that students produce something of value. It is probably the most exciting part of a REAL because students work on projects and tasks that are relevant to them and to their peers. It keeps students busy and happy – or active – while helping them construct and evolve their knowledge structures.

Authentic learning contexts
The third attribute of REALs is that learning takes place within an authentic context. An authentic task, activity, or goal provide learning experiences as realistic as possible, taking into consideration the age and maturation level of the students and environmental constraints such as safety and resource availability.

An authentic context incorporates as much fidelity as possible to what students will encounter outside school in terms of tools, complexity, cognitive functioning, and interactions with people (Williams and Dodge, 1992). Therefore, creating an authentic learning context requires more than just presenting students with realistic problems or situations – it also means that students must address the problems or situations realistically as well (Honebein *et al*, 1993).

Authenticity is important to REALs for three reasons. First, realistic problems hold more relevance to students' needs and experiences because they can relate what they are learning to problems and goals that they see every day. Therefore, it encourages students to take ownership of the situation and their own learning. Second, because the situations students encounter during learning are authentic and reflect the true nature of problems in the real world, it develops deeper and richer (indexicalized and conditioned) knowledge structures, leading to a higher likelihood of transfer to novel situations. Finally, because complex problems require a team approach that provides natural opportunities for learners to test and refine their ideas and to help each other understand the content, it encourages collaboration and negotiation.

REAL strategy: anchored instruction
One of the ways to create authentic instruction in a REAL is to anchor that instruction in a realistic event, problem, or theme (CTGV, 1990, 1992a, 1992b, 1993a, 1993d). Anchored instruction is fixed within a real-world event that is appealing and meaningful to students (Bransford *et al*, 1990) and involves complex contexts that require students to solve interconnected sub-problems. Because students are encouraged to work together to solve these complex problems, they are exposed to multiple perspectives in an environment that gives them an opportunity to test out their ideas, solutions, and processes (CTGV, 1992b).

At the heart of the model is an emphasis on the importance of creating an anchor or focus to generate interest, and to enable students to identify and solve problems and pay attention to their own perception and comprehension of these problems' (Bransford *et al*, 1990, p. 123).

In anchored learning situations, students develop component skills and objectives in the context of meaningful, realistic problems and problem-solving activities. These complex contexts are called 'macrocontexts' (Williams and Dodge, 1992, p. 373). Addressing a key characteristic of constructivism – indexed knowledge acquisition – the primary goal of anchored instruction (and REALs) is to overcome the problem of inert knowledge. For example, students in an instructional design and development class work in teams with actual clients to develop instruction that will be delivered to another group of students. They must define the problem, identify resources, set priorities, and explore alternative solutions with the same skills and abilities that are required during realistic, outside-the-classroom problem-solving and decision-making activities. This is in direct contrast to the way students develop component skills and objectives in a more conventional classroom environment by working simplified, compartmentalized, and decontextualized problems. Simply stated, it is the difference between providing meaningful, authentic learning activities and 'I'm never going to use this' activities.

Anchored instruction shares many features of programmes that are case-based and problem-based (Barrows, 1985; Spiro *et al*, 1991; Williams and Dodge, 1992). The idea is to let learners experience the intellectual changes that experts feel when modifying their own understandings from working with realistic situations (CTGV, 1992b).

Effective anchors are intrinsically interesting, fostering ownership, and helping students notice the features of problem situations that make particular actions relevant (Bransford *et al*, 1990). The CTGV (1991) uses the following design principles when creating anchored instruction. First, they use a video-based presentation format because of the dramatic power of the medium and because of the use of multiple modalities, realistic imagery, and omnipresence in our culture. Second they present a problem using actors and a narrative format for interest. Third, the problem solution requires a generative learning format in which students must identify pertinent information in the fourth feature, embedded data design. Fifth, the problem is complex, with the possibility of multiple solutions, and requires a team approach. Sixth, they use pairs of similar problems in different contexts to enrich the indexicalization of knowledge structures. Finally, they attempt to draw links across the curriculum to enhance the relevance of the problem.

One of the CTGV's projects in anchored instruction is the *Jasper Woodbury* series (CTGV, 1992b). Jasper is a video-based series designed to promote problem-posing, problem-solving, reasoning, and effective communication. Each of Jasper's adventures is a 15- to 20-minute story in which the characters encounter a problem that the students in the classroom must solve before they are allowed to see how the movie characters solved the problem. The Jasper series helps students learn to break a problem into parts, generate sub-goals, find and identify relevant information, generate and test hypotheses, and co-operate with others.

REAL strategy: cognitive flexibility theory
Cognitive Flexibility Theory (CFT) also addresses the need for authentic learning contexts in order to help learners develop conditionalized and indexicalized knowledge structures (Spiro *et al*, 1991; Jacobson and Spiro, 1992).

Essentially, the theory states that cognitive flexibility is needed in order to construct an ensemble of conceptual and case representations necessary to understand a particular problem-solving situation. The idea is that we cannot be said to have a full understanding of a domain unless we have the opportunity to see different case representations (Borsook and Higginbotham-Wheat, 1992, p. 63).

CFT attempts to teach content in ill-structured domains, that is, in domains where the knowledge-base is so vast and complex that multiple solutions to problems are possible and likely. There are no clear-cut answers in ill-structured domains, so simple algorithms often fail. Ill-structured domains include law, medicine, and education. So, CFT emphasizes the following instructional strategies to help learners develop rich and deep knowledge structures (Jacobson, 1994):

- CFT uses several cases and rich examples in their full complexity. One of the tenets of CFT is to avoid over-simplifying knowledge and examples because this leads to future misunderstandings that are difficult to change.

- CFT uses multiple forms of knowledge representation, providing examples in several kinds of media. CFT encourages students to look at knowledge in several ways and from several perspectives.

- CFT links abstract concepts to case examples and brings out the generalizable concepts and strategies applicable to other problems or cases.

- To avoid the mistakes of over-simplification, CFT presents a number of examples to make apparent, rather than hide, the variability of concepts and themes within the domain.

Although authentic learning contexts help students to develop knowledge that can be transferred and applied to new problems and situations, the fact that students are engaged in authentic activities creates some unique assessment problems. These issues will be examined next.

Authentic assessment strategies

Going hand in hand with the need to develop authentic learning contexts, the fourth REAL attribute is the use of authentic assessment strategies to evaluate student performance. Conventional schooling relies on standardized and paper/pencil tests to measure the quantity of knowledge that students have accrued. But traditional tests, written reports, and grading schemes are inappropriate measures (Frederiksen and Collins, 1989), time-consuming to administer and score (Williams and Dodge, 1992), and poor indicators of how students will perform in actual problem-solving conditions. Williams and Dodge also state that students are often assessed on skills different from ones that are taught, and experience problem-solving assessments that tend to be subjective. Testing and assessment must recognize the importance of the organization of the knowledge-base and its connectedness to contexts.

Wiggins (1989) contrasts authentic tests, which he describes as contextualized, complex intellectual challenges against multiple-choice measures that he describes as fragmented and static. According to Wiggins, authentic tests include the following criteria:

- The intellectual design features of tests and evaluation tasks must emphasize realistic complexity, stress depth more than breadth, include ill-structured tasks or problems, and require students to contextualize content knowledge.

- Standards of grading and scoring features should include complex multi-faceted criteria that can be specified and that are reliable across multiple scorers. What constitutes a high level of performance should be explainable to students and teachers *before* they take the test. Teachers often claim that criteria are subjective, but this is seldom the case; most criteria can be described with some thoughtful effort.

- Tests and evaluations must be diverse, and must recognize the existence of multiple kinds of intelligences. In terms of fairness and equity, evaluations and assessments should allow students to use their strengths within areas where their interests lie.

REAL example: learning in design

Carver *et al* (1992) elaborate on this theme by proposing an extensive list of behaviours needed by students in a REAL. In their particular manifestation of a REAL, they consider the classroom a design community in which students design instruction for other students, documentaries for local media, and other exhibits for the community. Their program has the same goals of high-level thinking, reflection, and transfer as the other REAL strategies described:

> The instructional virtues of these design experiences include the opportunity to develop and coordinate a variety of complex mental skills, such as decomposing a topic into subtopics, gathering data from a variety of sources, organizing diverse and often contradictory information, formulating a point of view, translating ideas into a presentation targeted at a particular audience, evaluating the design, and making revisions based on the evaluations. (Carver *et al*, 1992, p. 386)

Again, there are several parallels with the other examples that we have discussed. Their REAL focuses on complex mental skills; analysing, comparing, and manipulating information; working on authentic, community-based tasks; and working with others.

To evaluate fairly students working in this environment, teachers need a clear specification of the skills students need for design tasks and prescriptions for how teachers can effectively support those skills (see also Agnew *et al*, 1992). Specification of skills and prescriptions of support are two parts of assessment that must be linked for fair assessment. If a skill cannot be supported by the teacher or some kind of scaffolding technique, then it cannot be fairly evaluated. It may, in fact, be outside the ZPD (Zone of Proximal Development) and beyond the current capability of the student. One of the teacher's jobs in a REAL is to specify skills and performances that can be supported so the student can grow in ability. Carver *et al* (1992) break the important behaviours for their environment into:

- project-management skills, including creating a timeline, allocating resources, and assigning team roles;

- research skills, including determining the nature of the problem, posing questions, searching for information, developing new information, and analysing and interpreting information;

- organization and representation skills, including choosing the organization and structure of information, developing representations (text, audio, and graphics), arranging structure and sequence; and juggling constraints;

- presentation skills, including transferring their design into media and arousing and maintaining audience interest; and

- reflection skills, including evaluating the process and revising the design.

Their criteria work for their learning environment. Other models may need to revise some of the specifics, though the five main categories provide an excellent starting place in specifying skills targeted for assessment in authentic assessments. Goldman, Pellegrino, and Bransford's (1994) discussion of assessment in the CTGV *Jasper* series suggests the following assessment areas: assessment of complex mathematical problem-solving (mathematics is the content domain of *Jasper*), measures of group problem-solving performance, assessment of extensions into other areas of the content area, and assessment of cross-curricular extensions. Although conducting an assessment for a REAL is more work than conventional assessment, it is also an integral part of the learning process rather than a periodic quantifiable measure. Authentic assessment provides feedback and information that is useful for planning future learning activities.

Assessment in REALs means that we have to consider more varied techniques. Neuman (1993), in her work with the Perseus hypermedia program, suggests several alternatives. First, she suggests that teachers use more observations including evaluator observations of performance processes, think-alouds by students, and automatic transaction monitoring. Second, she suggests using interviews of students, instructors, and staff using both questionnaires and focus groups. Finally, she suggests using document and product analysis including assignments, syllabi, essays, journals, paths, reports, documentation, and presentations.

Co-operative support

The fifth and final characteristic of REALs, and one that is an important feature of almost all the other REAL strategies described above, acknowledges the transactional nature of knowledge and suggests that a shift be made to focus on social practice, meaning and patterns (Roth, 1990). 'All cooperative learning methods share the idea that students work together to learn and are responsible for one another's learning as well as their own' (Slavin, 1991, p. 73). Working in peer groups helps students refine their knowledge through argumentation, structured controversy, and the sharing and testing of ideas and perspectives. Additionally, students are more willing to take on the extra risk required to tackle complex, ill-structured, authentic problems when they have the support of others in the co-operative group. Co-operative learning and problem-solving groups also address students' needs for scaffolding during unfamiliar learning and problem-solving activities; therefore, with the support of others in the group, students are more likely to achieve goals they may not have been able to meet on their own.

Constructivists argue that co-operative learning and problem-solving groups facilitate generative learning. Some of the generative activities that students engage in co-operative groups include (Brown *et al*, 1989):

- Collective problem-solving. Groups give rise synergistically to insights and solutions that would not come about individually.

- Displaying multiple roles. Group participation means that the members must understand many different roles. They also may play different roles within the group to gain additional insights.

- Confronting ineffective strategies and misconceptions. Teachers do not have enough time to hear what students are thinking or how they are thinking. Groups draw out, confront, and discuss both misconceptions and ineffective strategies.

- Providing collaborative work skills. Students learn to work together in a give-and-take interaction rather than just dividing the workload.

Research indicates that co-operative learning, when implemented properly, is highly successful. Slavin (1991) provides the following four summary statements regarding research findings in co-operative learning:

- Successful co-operative-learning strategies always incorporate the two key elements of group goals and individual accountability.

- When both group goals and individual accountability are used, achievement effects are consistently positive. Slavin's review found that 37 of 44 experimental/control comparisons of at least four weeks' duration found significantly positive effects for co-operative group methods with none favouring traditional methods.

- Positive achievement effects are present to about the same degree across all US grade levels (2–12), in all major subjects, and in urban, rural, and suburban schools. Effects are equally positive for high, average, and low achievers.

- Positive effects of co-operative learning are consistently found on such diverse outcomes as self-esteem, inter-group relations, acceptance of academically handi-capped students, attitudes toward school, and ability to work co-operatively.

REAL strategy: problem-based learning
Problem-based learning (PBL) embodies the definition of a rich environment for active learning, and incorporates all attributes previously described. PBL is 'the learning that results from the process of working toward the understanding or resolution of a problem' (Barrows and Tamblyn, 1980, p. 18). PBL found initial acceptance in the medical field, and has grown to become a major learning system for a number of medical, law, and business schools. It is also being adapted for use in secondary schools and corporate training environments (Dunlap, 1995). PBL reflects the REAL attribute that knowledge is constructed rather than received, for it is based on the assumption that knowledge arises from work with an authentic problem (Coltrane, 1993). Benor (1984) states that:

> Problem-based learning in the context of medical education means self-directed study by of [sic] learners who seek out information pertinent to either a real-life or a simulated problem. The students have to understand the problem to the extent that its constituents can be identified and defined. The learners have then to collect, integrate, synthesize and apply this information to the given problem, using strategies that will yield a solution. (p. 49)

How does PBL work? Savery and Duffy (1994) describe four characteristics of PBL.

First, PBL environments include the learning goals of realistic problem-solving behaviour, self-directed learning, content knowledge acquisition, and the development of metacognitive skills.

Second, Savery and Duffy state that PBLs are based on problems that are generated because they raise relevant concepts and principles that are authentic. Problems must be authentic because it is difficult to create artificial problems that maintain the complexity and dimensions of actual problems. Recall that we encountered the need for complexity in the REAL strategies of anchored instruction and cognitive flexibility theory. Realistic problems also have a motivational effect. They tend to engage learners more because they want to know the outcome of the problem. When the learning context is similar to the situation in which the learning is to be applied, learning transfer is more likely to occur. Therefore, we see the continual reference to the necessity for transfer in PBL.

Third, the actual presentation of the problem is a critical component of PBLs. Problems are encountered before any preparation or study has occurred (Barrows, 1980). The problem must be presented in a realistic way that encourages students to adopt and take ownership for the problem (Barrows and Tamblyn, 1980; Savery and Duffy, 1994). Work on the problem begins with activating prior knowledge to enable students to understand the structure of the new information. Learners state what they already know about the problem domain. They use that knowledge to form hypotheses or ideas about potential solutions. We also see this principle emphasized in intentional learning, because students must ask themselves what they know about a subject before creating learning plans. The data must be embedded in the problem presentation (refer back to the example of anchored instruction) but must not highlight the critical factors in the case. Students must make their own decisions about what is critical because that is cognitively authentic – it reflects actual job performance (Savery and Duffy, 1994).

Fourth, the facilitator has a crucial role comparable to the roles described in anchored instruction and reciprocal teaching. The facilitator interacts with the students at a metacognitive level, helping them ask the right questions and monitor their own progress. Facilitators avoid expressing opinion, giving information, or leading to a correct answer. Their role is to challenge the students, and help them reflect on what they are learning. (Savery and Duffy, 1994).

Co-operative learning is a critical component of PBL for it is used from the beginning to the end of the problem-solution process. Members of the group listen to the problem presentation together. They analyse the problem's components, recall what they know, hypothesize, consider possible resources, and choose directions to go. They test and help each other. They work together on the solutions and reach consensus on final actions. The entire process from beginning to end is co-operative. Co-operative learning is also used for its motivational factors.

Problem discussion also increases motivation by gaining and maintaining student interest (attention), by relating the learning to student needs or helping students to meet personal goals (relevance), by providing conditions conducive to student success (confidence), and through the motivation provided by that mastery of the task(s) (satisfaction) (Coltrane, 1993, pp. 12–13).

PBL is the epitome of the REAL constructive learning process. Students work with problems in a manner that fosters reasoning and knowledge application appropriate to their levels of learning. In the process of working on the problem and with their peers, students identify areas of learning to guide their own individualized study. The skills and knowledge acquired by this study are applied back to the problem to evaluate the effectiveness of learning and to reinforce learning. The learning that has occurred in work with the problem and in individualized study is summarized and integrated into the student's existing knowledge structure.

Conclusion

Times have changed. People now need to be able to think flexibly and creatively, solve problems, and make decisions within complex, ill-structured environments. Given these changes, our assumptions on learning and education are out-dated, forcing us to modify our assumptions based on current theoretical views of learning. These new assumptions, supported by the theories and philosophies of constructivism, require different instructional methods, techniques, and strategies than have been conventionally used in classroom settings. Reflecting a constructivist view of learning, REALs provide a way for us to address these new assumptions in order to meet the educational demands of a changing society.

We have looked at each of the five main attributes of REALs that support the goals of constructivism: (1) student responsibility and initiative, (2) generative learning activities, (3) authentic learning contexts, (4) authentic assessment strategies, and (5) co-operative support. Each REAL attribute builds upon and uses the others. None of the attributes are mutually exclusive, and no one attribute is more important than another; you cannot implement one feature without incorporating the others to some degree. In effect, the attributes of REALs mirror the comprehensive, integrated, and holistic nature of REALs. The characteristics are symbiotic, with one feature both supporting and needing the others to create a successful rich environment for active learning.

References

Adams, L., Kasserman, J., Yearwood, A., Perfetto, G., Bransford, J. and Franks, J. (1988), 'The effects of facts versus problem-oriented acquisition', *Memory and Cognition*, 16, 167–75.

Agnew, P., Kellerman, A., and Meyer, J. (1992), 'Constructing multimedia: solutions for education', paper presented at the 34th Annual International Conference of the Association for the Development of Computer-Based Instructional Systems, Norfolk VA.

Albanese, M. A. and Mitchell, S. (1993), 'Problem-based learning: a review of literature on its outcomes and implementation issues', *Academic Medicine*, 68 (1), 52–81.

Allinson, L. and Hammond, N. (1990), 'Learning support environments: rationale and evaluation', *Computers in Education*, 15 (1), 137–43.

Ambruso, D. and The Transfusion Medicine Group (1994), *Transfusion Medicine Modules* [computer programs], Denver CO, The Bonfils Blood Center.

American Association for the Advancement of Science (1989), a project 2061 report on literacy goals in science, mathematics, and technology, Washington DC, AAAS.

Barrows, H. S. (1985), *How to Design a Problem-based Curriculum for the Preclinical Years*, New York, Springer-Verlag.

Barrows, H. S. and Tamblyn, R. M. (1980), *Problem-based Learning: An Approach to Medical Education*, New York, Springer Publishing Company.

Bednar, A. K., Cunningham, D., Duffy, T. M. and Perry, J. D. (1991), 'Theory into practice: how do we link?', in Anglin, G. J. (ed.), *Instructional Technology: Past, Present, and Future* (pp. 88–101), Englewood CO, Libraries Unlimited.

Benor, D. E. (1984), 'An alternative, non-Brunerian approach to problem-based learning', in Schmidt, H. G. and d. Volder, M. L. (eds), *Tutorials in Problem-based Learning: New Directions in Training for the Health Professions* (pp. 48–58), Assen/Maastricht, Van Gor Cum.

Bereiter, C. and Scardamalia, M. (1989), 'Intentional learning as a goal of instruction', in Resnick, L. B. (ed.), *Knowing, Learning, and Instruction: Essays in Honor of Robert Glaser* (pp. 361–92), Hillsdale NJ, Lawrence Erlbaum.

Berryman, S. E. (1991), *Designing Effective Learning Environments: Cognitive Apprenticeship Models*, ERIC Document 337 689, 1–5.

Biggs, J. B. (1985), 'The role of metalearning in study processes', *British Journal of Educational Psychology*, 55, 185–212.

Blakey, E. and Spence, S. (1990), *Developing Metacognition*, ERIC Document 327 218, 1–4.

Borsook, T. K. and Higginbotham-Wheat, N. (1992), 'The psychology of hypermedia: a conceptual framework for R and D', paper presented at the 1992 National Convention of the Association for Educational Communications and Technology, Washington DC.

Bransford, J. D., Sherwood, R. D., Hasselbring, T. S., Kinzer, C. K. and Williams, S. M. (1990), 'Anchored instruction: why we need it and how technology can help', in Nix, D. and Spiro, R. (eds), *Cognition, Education, and Multimedia: Exploring Ideas in High Technology* (pp. 115–41), Hillsdale NJ, Lawrence Erlbaum.

Bransford, J. and Vye, N. J. (1989), 'A perspective on cognitive research and its implications for instruction', in Resnick, L. and Klopfer, L. E. (eds), *Toward the Thinking Curriculum: Current Cognitive Research* (pp. 173–205), Alexandria VA, ASCD.

Bransford, J., Goldman, S. R. and Vye, N. J. (1991), 'Making a difference in peoples' abilities to think: reflections on a decade of work and some hopes for the future', in Okagaki, L. and Sternberg, R. J. (eds), *Directors of Development: Influences on Children* (pp. 147–80), Hillsdale NJ, Lawrence Erlbaum.

Brown, A. L. (1978), 'Knowing when, where, and how to remember: a problem of metacognition', in Glaser, R. (ed.), *Advances in Instructional Psychology*, Hillsdale NJ, Lawrence Erlbaum.

Brown, A. L., Bransford, J. D., Ferrara, R. A. and Campione, J. C. (1983), 'Learning, remembering, and understanding', in Flavell, J. H. and Markman, E. M. (eds), vol. 3, *Handbook of Child Psychology: Cognitive Development* (pp. 177–266), New York, Wiley.

Brown, J. S., Collins, A. and Duguid, P. (1989), 'Situated cognition and the culture of learning', *Educational Researcher*, January–February, 32–42.

Bruner, J. S. (1961), 'The act of discovery', *Harvard Educational Review*, 21–32.

Butterfield, E. and Nelson, G. (1989), 'Theory and practice of teaching for transfer', *Educational Technology Research and Development*, 37 (3), 5–38.

Carver, S. M., Leherer, R., Connell, T. and Erickson, J. (1992), 'Learning by hypermedia design: issues of assessment and implementation', *Educational Psychologist*, 27 (3), 385–404.

Clark, R. E. (1994), 'Media will never influence learning', *Educational Technology Research and Development*, 42 (2), 21–9.

Clark, R. E. and Voogel, A. (1985). 'Transfer of training principles for instructional design', *Educational Communication and Technology Journal*, 33 (2), 113–25.

Clement, J. (1982). 'Algebra word problem solutions: thought processes underlying a common misconception', *Journal of Research in Mathematics Education*, 13, 16–30.

CTGV (Cognition and Technology Group at Vanderbilt) (1990), 'Anchored instruction and its relationship to situated cognition', *Educational Researcher*, 19 (6), 2–10.

CTGV (Cognition and Technology Group at Vanderbilt) (1991), 'Technology and the design of generative learning environments', *Educational Technology*, 31, 34–40.

CTGV (Cognition and Technology Group at Vanderbilt) (1992a), 'Anchored instruction in science and mathematics: theoretical basis, developmental projects, and initial research findings', in Duschl, R. A. and Hamilton, R. J. (eds), *Philosophy of Science, Cognitive Psychology, and Educational Theory and Practice* (pp. 244–73), New York, SUNY Press.

CTGV (Cognition and Technology Group at Vanderbilt) (1992b), 'The *Jasper Series* as an example of anchored instruction: theory, program description, and assessment data', *Educational Psychologist*, 27 (3), 291–315.

CTGV (Cognition and Technology Group at Vanderbilt) (1993a), 'Anchored instruction and situated cognition revisited', *Educational Technology*, 13 (3), 52–70.

CTGV (Cognition and Technology Group at Vanderbilt) (1993b), 'Designing learning environments that support thinking', in Duffy, T. M., Lowyck, J. and Jonassen, D. H. (eds), *Designing Environments for Constructive Learning* (pp. 9–36), New York, Springer-Verlag.

CTGV (Cognition and Technology Group at Vanderbilt) (1993c), 'Integrated media: toward a theoretical framework for utilizing their potential', *Journal of Special Education Technology*, 12 (2), 76–89.

CTGV (Cognition and Technology Group at Vanderbilt) (1993d), 'The *Jasper Series*: theoretical foundations and data on problem solving and transfer', in Penner, L. A., Batsche, G. M., Knoff, H. M. and Nelson, D. L. (eds), *The Challenges in Mathematics and Science Education: Psychology's Response* (pp. 113–52), Washington DC, American Psychological Association.

Collins, A. (1995), 'Learning communities', presentation at the annual conference for the American Educational Research Association, San Francisco CA, April, 1995.

Collins, A., Brown, J. S. and Holum, A. (1991), 'Cognitive apprenticeship: making thinking visible', *American Educator* (Winter), 6–11, 38–46.

Coltrane, L. (1993), 'An overview of problem-based learning in medical education' (class paper).

Dewey, J. (1910), *How We Think*, Boston, Heath.

Duffy, T. M. (1992), *The Strategic Teaching Framework*, Bloomington, IN.

Duffy, T. M. (in press), 'Strategic teaching framework: an instructional model for learning complex, interactive skills' in Dill, C. and Romiszowski, A. (eds), *Encyclopedia of Educational Technology*, Englewood NJ, Educational Technology Press.

Duffy, T. M. and Bednar, A. K. (1991), 'Attempting to come to grips with alternative perspectives', *Educational Technology*, 31 (9), 12–15.

Dunlap, J. C. (1995), 'Using constructivist training environments to meet long-term strategic training needs', paper presented at the Annual Conference of the Association for Educational Communications and Technology, Anaheim CA.

Dunlap, J. C. and Grabinger, R. S. (1992), 'Designing computer-supported intentional learning environments', paper presented at the Annual Conference of the Association for the Development of Computer-Based Instructional Systems, Norfolk VA.

Dunlap, J. C. and Grabinger, R. S. (1993), 'Computer-supported intentional learning environments: definition and examples', paper presented at the Annual Conference of the Association for Educational Communications and Technology, New Orleans LA.

Farnham-Diggory, S. (1992), *Cognitive Processes in Education* (second edition), New York, Harper Collins.

Feuerstein, R. (1979), *Instrumental Enrichment*, Baltimore MD, University Park.

Forman, G. and Pufall, P. (eds) (1988), *Constructivism in the Computer Age*, Hillsdale NJ, Lawrence Erlbaum.

Fosnot, C. (1989), *Inquiring Teachers, Inquiring Learners: A Constructivist Approach for Teaching*, New York, Teacher's College Press.

Frederiksen, J. R. and Collins, A. (1989), 'A systems approach to educational testing', *Educational Researcher*, 18, 27–32.

Gayeski, D. M. (ed.) (1995), *Designing Communication and Learning Environments*, Englewood Cliffs NJ, Educational Technology Publications.

Goldman, S. R., Pellegrino, J. W. and Bransford, J. (1994), 'Assessing programs that invite thinking', in Baker, E. and O'Neill, H. F. J. (eds), *Technology Assessment in Education and Training* (pp. x–y), Hillsdale NJ, Lawrence Erlbaum.

Goldman, S. R., Petrosino, A., Sherwood, R. D., Garrison, S., Hickey, D., Bransford, J. D. and Pellegrino, J.W. (1992), 'Multimedia environments for enhancing science instruction', paper presented at the NATO Advanced Study Institute on Psychological and Educational Foundations of Technology-Based Learning Environments, Kolymbari, Greece.

Goodman, N. (1984), *Of Mind and Other Matters*, Cambridge MA, Harvard University Press.

Grabinger, R. S. and Dunlap, J. C. (1994a), 'Implementing rich environments for active learning: a case study', paper presented at the Annual Conference of the Association for Communications and Technology, Nashville LA.

Grabinger, R. S. and Dunlap, J. C. (1994b), 'Technology support for rich environments for active learning', paper presented at the Annual Conference of the Association for Communications and Technology, Nashville LA.

Gurney, B. (1989), *Constructivism and Professional Development: A Stereoscopic View*, ERIC Document ED 305 259, 1–28.

Hannafin, M. J. (1992), 'Emerging technologies, ISD, and learning environments: critical perspectives', *Educational Technology Research and Development*, 40 (1), 49–63.

Honebein, P. C., Duffy, T. M. and Fishman, B. J. (1993), 'Constructivism and the design of learning environments: context and authentic activities for learning, in Duffy, T. M., Lowych, J. and Jonassen, D. H. (eds), *Designing Environments for Constructive Learning*, Hillsdale NJ, Lawrence Erlbaum.

Jacobson, M. J. (1994), 'Issues in hypertext and hypermedia research: toward a framework for linking theory-to-design', *Journal of Educational Multimedia and Hypermedia*, 3 (2), 141–54.

Jacobson, M. J. and Spiro, R. J. (1991), 'Hypertext learning environments and cognitive flexibility: characteristics promoting the transfer of complex knowledge', paper presented at the International Conference on the Learning Sciences, Evanston IL.

Jacobson, M. J. and Spiro, R. J. (1992), 'Hypertext learning environments, cognitive flexibility, and the transfer of complex knowledge: an empirical investigation', paper presented at the Annual Meeting of the American Educational Research Association, San Francisco CA.

Jonassen, D. H. (1994b), 'Sometimes media influence learning', *Educational Technology Research and Development*, 42 (2).

Jonassen, D. H. (1994c), 'Thinking technology: toward a constructivist design model', *Educational Technology*, 34 (3), 34–7.

Kozma, R. (1994), 'Media attributes', *Educational Technology Research and Development*, 42 (2).

Lebow, D. (1993), 'Constructivist values for instructional systems design: five principles toward a new mindset', *Educational Technology Research and Development*, 41 (3), 4–16.

Linn, M. C. (1986), *Establishing a Research Base for Science Education: Challenges, Trends, and Recommendations*, report of a National Science Foundation national conference, University of California.

Lockhart, R. S., Lamon, M. and Gick, M. L. (1988), 'Conceptual transfer in simple insight problems', *Memory and Cognition*, 16, 36–44.

Lynton, E. (1989), *Higher Education and American Competitiveness*, National Center on Education and the Economy.

Lynton, E. and Elman, S. (1987), *New Priorities for the University*, San Francisco CA, Jossey-Bass.

Mann, L. (1979), *On the Trail of Process: A Historical Perspective on Cognitive Processes and their Training*, New York, Grune and Stratton.

Minstrell, J. A. (1989), 'Teaching science for understanding', in Resnick, L. B. and Klopfer, L. E. (eds), *Toward the Thinking Curriculum: Current Cognitive Research* (pp. 129–49), Alexandria VA, ASCD.

Morris, C. D., Bransford, J. D. and Franks, J. J. (1979), 'Levels of processing versus transfer appropriate processing', *Journal of Verbal Learning and Verbal Behavior*, 16, 519–33.

National Council of Teachers of Mathematics (1989), *Curriculum and Evaluation Standards for School Mathematics*, Reston VA, NCTM.

Neuman, D. (1993), 'Evaluation of the Perseus project', paper presented at the 1993 National Conference of the Association for Educational Communications and Technology, New Orleans LA.

Nickerson, R. S. (1988), 'On improving thinking through instruction', *Review of Research in Education*, 15, 3–57.

Palincsar, A. S. (1990), 'Providing the context for intentional learning', *Remedial and Special Education*, 11 (6), 36–9.

Palincsar, A. S. and Klenk, L. (1992), 'Fostering literacy learning in supportive contexts', *Journal of Learning Disabilities*, 25 (4), 211–25.

Perelman, L. J. (1992), 'Living in the gap between old and new: managing transitions', paper presented at the Technology in Education Conference, Steamboat Springs CO.

Perfetto, B. A., Bransford, J. D. and Franks, J. J. (1983), 'Constraints on access in a problem solving context', *Memory and Cognition*, 11, 24–31.

Perkins, D. N. (1991), 'What constructivism demands of the learner', *Educational Technology*, 31 (9), 19–21.

Resnick, L. (1987), *Education and Learning to Think*, Washington DC, National Academy Press.

Resnick, L. B. and Klopfer, L. E. (eds) (1989), *Toward the Thinking Curriculum: Current Cognitive Research*, Alexandria VA, ASCD.

Ridley, D. S., Schutz, P. A., Glanz, R. S. and Weinstein, C. E. (1992), 'Self-regulated learning: the interactive influence of metacognitive awareness and goal-setting', *Journal of Experimental Education*, 60 (4), 293–306.

Robertson, W. C. (1990), 'Detection of cognitive structure with protocol data: predicting performance on physics transfer problems', *Cognitive Science*, 14, 253–80.

Roth, W.-M. (1990), *Collaboration and Constructivism in the Science Classroom*, ERIC Document 318 631, 1–39.

Savery, J. R. and Duffy, T. M. (1994), 'Problem based learning: an instructional model and its constructivist framework', *Educational Technology* (August).

Scardamalia, M. and Bereiter, C. (1991), 'Higher levels of agency for children in knowledge building: a challenge for the design of new knowledge media', *The Journal of the Learning Sciences*, 1 (1), 37–68.

Scardamalia, M., Bereiter, C., McLean, R. S., Swallow, J. and Woodruff, E. (1989), 'Computer-supported intentional learning environments', *Journal of Educational Computing Research*, 5 (1), 51–68.

Schoenfeld, A. H. (1989), 'Teaching mathematical thinking and problem solving', in Resnick, L. B. and Klopfer, L. E. (eds), *Toward the Thinking Curriculum: Current Cognitive Research* (pp. 83–103), Alexandria VA, ASCD.

Segal, J., Chipman, S. and Glaser, R. (eds) (1985), *Thinking and Learning Skills: Relating Instruction to Basic Research*, vol. 1, Hillsdale NJ, Lawrence Erlbaum.

Shank, R. C. (1990), 'Case-based teaching: four experiences in educational software design', *Interactive Learning Environments*, 1 (4), 231–53.

Slavin, R. E. (1991), 'Synthesis of research on cooperative learning', *Educational Leadership*, 48 (5), 71–82.

Spiro, R. J., Feltovich, P. L., Jacobson, M. J. and Coulson, R. L. (1991), 'Cognitive flexibility, constructivism, and hypertext: random access instruction for advanced knowledge acquisition in ill-structured domains', *Educational Technology*, 31 (5), 24–33.

Tulving, E. and Thompson, D. M. (1973), 'Encoding specificity and retrieval processes in episodic memory', *Psychological Review*, 80, 352–73.

Von Wright, J. (1992), 'Reflections on reflection', *Learning and Instruction*, 2, 59–68.

Vygotsky, L. S. (1978), *Mind in Society*, Cambridge MA, Harvard University Press.

Weinstein, C. E., Goetz, E. T. and Alexander, P. A. (1988), *Learning and Study Strategies: Issues in Assessment, Instruction, and Evaluation*, San Diego CA, Academic Press.

Wheatley, M. (1992), *Leadership and the New Science*, San Francisco CA, Berrett-Koehler.

Whitehead, A. N. (1929), *The Aims of Education and Other Essays*, New York, Macmillan.

Wiggins, G. (1989), 'A true test: toward more authentic and equitable assessment', *Phi Delta Kappan*, 70, 703–13.

Williams, M. D. and Dodge, B. J. (1992), 'Tracking and analyzing learner-computer interaction', paper presented at the 1992 National Conference of the Association for Educational Communications and Technology, New Orleans LA.

Williams, S. M. (1992), 'Putting case-based instruction into context: examples from legal and medical education', *The Journal of the Learning Sciences*, 2 (4), 367–427.

Update – Rich environments for active learning

In the five years since this article was published, our belief in the importance of lifelong learning has increased because of an emerging emphasis on Web-based instruction as a mechanism for the delivery of professional-development opportunities. Web-based instruction enables greater individualization and flexibility, creating an increased demand for self-directed learning.

Using Web-based learning opportunities to meet professional-development goals requires well-developed lifelong learning skills and strategies, such as goal-setting, action-planning, learning-strategy selection and assessment, resource selection and evaluation, reflective learning and time management. Furthermore, one of the problems associated with Web-based, individualized environments is a sense of isolation due to lack of interaction (learner-to-facilitator, learner-to-learner). Interaction is a critical component of constructivist learning environments, whether done on the Web or in person, because learning occurs in a social context through collaboration, negotiation, debate, peer review and mentoring.

Recent work has thus focused on the importance of interaction in Web-based learning environments. From a constructivist viewpoint, studies on Web-based learning environments have shown that there are three critical components to interaction. First, an academic (learner-to-content) component occurs when learners access online materials and receive task-orientated feedback from the facilitator or from a technology-driven feedback system. Second, a collaborative (learner-to-learner) component occurs when learners are engaged in discourse, problem-solving, and product-building using Web-mediated communication and collaboration tools. This integration component helps learners validate their learning experiences, and requires a level of reflective articulation that promotes collective knowledge-building and a deeper personal understanding of what is being studied. Finally, an interpersonal/social component occurs when learners receive feedback from the facilitator or peers and colleagues in the form of personal encouragement and motivational assistance. Social interaction can contribute to learner satisfaction and frequency of interaction in an online learning environment. Without the opportunity actively to interact and exchange ideas with each other and the facilitator, learners' social, as well as cognitive, involvement in the learning environment is diminished.

Five years ago, we probably would have said that all of the attributes of REALs were equally important. Today, we believe that meaningful interaction is the foundation for making REALs work. We have found that the REAL guidelines have helped us develop strong interactive online learning experiences that address the problem of isolation. The holistic approach of REALs provides an excellent framework for Web-based learning environments.

Contact author

Scott Grabinger is an associate professor at the University of Colorado at Denver. He is currently on special assignment directing the Technology and Learning Team and Faculty Technology Studio. Scott also teaches courses in the Information and Learning Technologies Program including message design, student-centred learning environments, and doctoral seminars. [*sgrabing@carbon.cudenver.edu*]

Learning relationships from theory to design

C. J. H. Fowler*[1] and J. T. Mayes**
*Advanced Communications Research, BT Adastral Park
**Learning and Educational Development, Glasgow Caledonian University

Initially published in 1999

This paper attempts to bridge the psychological and anthropological views of situated learning by focusing on the concept of a learning relationship, and by exploiting this concept in our framework for the design of learning technology. We employ Wenger's (1998) concept of communities of practice to give emphasis to social identification as a central aspect of learning, which should crucially influence our thinking about the design of learning environments. We describe learning relationships in terms of form (one-to-one, one-to-many etc.), nature (explorative, formative and comparative), distance (first-, second-order), and context, and we describe a first attempt at an empirical approach to their identification and measurement.

Introduction

Over the last five years we have seen a very significant increase in the use of Information Communication Technologies (ICT) in schools, colleges and university. For example in 1998, there were over 195 accredited US universities offering a thousand or more distance learning courses (Philips and Yager, 1998). By no means were all of these new courses associated with educational innovation. The speed and ease of implementation of Web-based approaches, in particular, is resulting in design by imitation of current courses and methods, with a real lack of innovation or utilization of the power inherent in technology-based learning. Although matters are improving (see for example Brown, 1999), part of the reason for this failure to innovate is, we argue, because of the large gap between theory and practice.

This paper sets out to bridge what might be regarded as the psychological and anthropological views of situated learning by focusing on the concept of *learning relationships*. The aim is to consider how, in designing learning environments and tasks, we might usefully shift our focus away from the design of activities and examine more carefully what it is that motivates learners to engage in the learning task in the first place. What is it that would encourage them to strive for mastery? We ask what it is that the individual learner characteristically brings to the activity and to the environment, beyond the rather unproven variables of learning style. Finally we consider, with an example from a current project, how the concept of learning relationships might be elaborated empirically and used in the design of learning technology.

The theoretical underpinning of attempts to design learning environments has recently witnessed two distinct shifts of emphasis. First, there has been a shift from a representational view of learning in which an acquisition metaphor guided design (e.g. Anderson *et al*, 1990), to a constructivist view in which learning is primarily developed through activity (Papert, 1990). Brown *et al* (1989) argued that we should consider concepts as tools, to be understood through use, rather than as self-contained entities to be delivered through instruction. A second shift, however, has been away from a focus on the individual, towards a new emphasis on social contexts for learning (Glaser, 1990). There are at least two aspects to this. First, we have seen for about fifteen years a new emphasis on learning through collaboration and co-operation (see Kaye, 1992), but, even more importantly perhaps, came the notion of situated learning (Lave, 1988; Lave and Wenger, 1991; Suchman 1987).

As Barab and Duffy (1999) point out, there are at least two 'flavours' to situated learning. One, following Resnick's (1987) articulation of the nature of informal, out-of-school learning, and Cole and Scribner's (1974) approach to the cognitive implications of formal and informal learning, can be regarded as a socio-psychological view of situativity. They emphasized the importance of context-dependent learning in informal settings. Children in traditional societies learn by observing others, a form of apprenticeship where the knowledge often remains tacit. In contrast, formal education removes the child from context by locating learning in schools. The separation of where you learn from what you learn can be seen as playing an important part in developing abstract thinking in the child, but a growing dissatisfaction with formal schooling has led to a 'participation' metaphor replacing the 'acquisition' metaphor (Sfard, 1998). See Laurillard (1993) for a discussion of the 'abstraction' issue in the context of university education.

This activity-based view of situated learning led to the design of what Barab and Duffy (following Senge, 1994) call 'practice fields'. These represent constructivist tasks in which every effort is made to make the learning activity authentic to the social context in which the skills or knowledge are normally embedded. Examples of approaches to the design of practice fields are problem-based learning (Savery and Duffy, 1996), anchored instruction (CTGV, 1993) and cognitive apprenticeship (Collins *et al*, 1989). Here, the main design emphasis is on the relationship between the nature of the learning task in educational or training environments, and its characteristics when situated in real use. This connection between the classroom and the real world through 'practice fields' has certain cognitive implications. It is unclear, for example, how children abstract out general principles and avoid becoming 'context-bound' (cf. Cole and Scribner, 1974).

The second view of situativity, however, is a social anthropological one in which the influence of a wider social context is emphasized (Lave and Wenger, 1991). Here we find the concept of a community of practice introduced. With it comes an emphasis on the practitioner's relationship with a wider but identifiable group of people rather than the relationship of the activity itself to the wider practice, even though it is the practice itself that identifies the community. This provides a different perspective on what is 'situated'. For Lave (1997) and Wenger (1998) it is not just the meaning to be attached to an activity that is derived from a community of practice, but a much more fundamental aspect. The individual's *identity* is shaped by the relationship. Lave and Wenger's perspective, however, emphasizes the stable and long-term nature of communities of practice. In our view this restricts the potential usefulness of the idea for the design of learning environments, where short-term and more fragile groups may nevertheless exert a powerful influence on the motivation to learn. Indeed, the social psychological literature on social identification (Turner, 1991; Hogg and Adams, 1988) seems to demonstrate that temporary but real social identities can be created through group membership.

The starting point for this paper is a description of two conceptual frameworks. Wenger (1998) describes how communities of practice underpin learning in both formal and informal contexts. Mayes and Fowler (1999) map a cognitive-constructivist account of learning onto the design of learning environments.

Wenger's conceptualization

For Wenger, knowledge is a matter of competence in a valued enterprise. The value is given by social participation – in particular, by being an active participant in the practices of social communities, and by constructing an *identity* in relation to each community. Participating in a community – it may be a project team, say, or a member of a professional group – is both a kind of action and a form of belonging. Wenger's social theory of learning, therefore, recognizes both strands of situativity described above: meaning is given both to the situated activities themselves and to the process of social identification which drives the learners' activity.

Wenger argues that the social production of meaning is the relevant unit of analysis for practice. Meaning is continually *negotiated* through the processes of *participation* and *reification*. It is worth quoting Wenger directly on *participation:*

> Participation is an active process, but I will reserve the term for actors who are members of social communities. For instance, I will not say that a computer 'participates' in a community of practice, even though it may be part of that process and play an active role in getting things done . . . In this regard what I take to characterise participation is the possibility of mutual recognition . . . What we recognise has to do with our mutual ability to negotiate meaning . . . The relations between parents and children, or between workers and supervisor, are mutual in the sense that participants shape each others' experiences of meaning.

For Wenger, participation is not the same thing as collaboration. It goes beyond direct engagement in specific activities, and is 'not something we turn on and off'. Wenger argues that our engagement with tasks is social, even where it does not directly involve any kind of explicit dialogue.

Being in a hotel room by yourself preparing a set of slides for a presentation the next morning may not seem like a particularly social event, yet its meaning is fundamentally social. Not only is the audience with you as you attempt to make your points understandable, but your colleagues are there too, looking over your shoulder, as it were, representing for you your sense of accountability to the professional standards of your community.

We can participate in global communities of practice, but more generally we tend to engage with local ones. How can we identify that a community of practice has formed? Wenger suggests that we must look for sustained mutual relationships, which will tend to involve shared ways of engaging in doing things together, and the rapid flow of information. The discourse in such a community will have the appearance of an ongoing process, with many shared assumptions and an absence of introductory preambles. There will be jargon, inside jokes, many communication shortcuts, and the rapid setup of a problem to be discussed.

For Wenger, there are three stages of coming to belong to a community of practice:

- *imagination*: through orientation and exploration, we identify with a community of practice;
- *engagement*: through participating in a community we value, we come to belong to that community;
- *alignment*: we connect to a new framework of convergence.

Wenger articulates the same basic point that this paper is arguing, namely that issues of education should be addressed first and foremost in terms of identities and modes of belonging, and only secondarily in terms of skills and information. This view encourages us to consider educational designs not just in terms of techniques for supporting the construction of knowledge (let alone in terms of delivery of a curriculum), but more generally in terms of their effects on the formation of identities. What does this mean for design? Wenger goes some way towards operationalizing his framework (although not very far) by proposing that students need:

- places of engagement;
- materials and experiences with which to build an identity;
- *ways of making their actions matter.*

The constructivist approach requires the design of tasks which are personally meaningful for learners. Wenger's conceptualization of communities of practice gives us a way of defining personal meaning in a way that is not just circular. However, it is not a description of learning *per se*, or of how people learn together. It provides a very high-level design heuristic and in that sense it tells us where we should start looking for design principles particularly within an organizational context.

A framework for the design of learning technology

When BT became interested in designing and building online educational services, it soon became apparent that it was hard to move from theoretical concepts, e.g. constructivism or conceptualization, to a set of design principles or guidelines that could help engineer the next generation of educational technology. We needed conceptual frameworks that bridged the theory and design. Mayes (1995) offered such a framework. This framework described three broad modes of learning and then mapped these onto appropriate design principles. The modes or stages of learning were:

- Conceptualization: the coming into contact with other people's concepts.

- Construction: the building and testing of one's knowledge through the performance of meaningful tasks.

- Dialogue: the debate and discussion that results in the creation of new concepts.

It is important to note that 'conceptualization' is about *other people's concepts,* 'construction' is about *building knowledge* from combining your own and other people's concepts into something meaningful. 'Dialogue' refers back to the creation of new concepts (rather than knowledge) that then triggers another cycle of the reconceptualization process.

Fowler and Mayes (1997) have extended the notion of dialogue to include dialogues or learning conversations for clarification and confirmation at the conceptualization stage, and dialogue for co-operation and collaboration at the construction stage. The framework proposes that each learning stage should be associated with a different kind of pedagogy, which in turn implies a different kind of learning environment and a different form of courseware. At conceptualization, the learner tends to rely on traditional forms of courseware which explain and illuminate the concepts in the subject matter, typically building an exposition through textbooks, multimedia or simulations. This type of courseware was referred to as *primary. Secondary* courseware took the form of materials or tools that supported 'learning by doing', while *tertiary* courseware referred to the captured dialogues which gave rise to significant learning episodes and the availability of this material to others in the support of vicarious learning (McKendree *et al,* 1998).

The framework can be developed, by elaborating the role of dialogue. We have previously emphasized the importance of dialogue in learning, elevating it above, or at least level with, conceptualization as a pedagogical design principle. Dialogue provides the vehicle for conceptual movement. It facilitates the transition between the stages and the advance from one reconceptualization cycle to the next.[2] To emphasize the central role of dialogue even further we now view it not as a separate stage, but as integral to the whole cycle. The third stage we have now labelled *'identification'*. At the identification stage, the learner has reached a sufficient level of understanding to be able to relate to other conceptualizations and thus begin the process again. By calling this process identification it is intended to emphasize our belief that it is social in origin, and can only be understood in terms of the learner's relationships with others. It makes the important design point that this stage cannot be fully understood by knowing all about the interactions between the subject matter, the tasks, or the pedagogical environment. The nature of the interactions between individual learners and other people now comes fully into focus.

Learning relationships: a new conceptualization

Wenger's account paints a rich picture of the social process of belonging and provides an explanation of personal motivation through identification with the practices of communities. People are not motivated to learn *per se*, but are motivated to join a community of practice – an end that can be best achieved through learning. It is this richer social context and explanation of motivation that are lacking in our previous framework.

The two frameworks impact on design in different ways. For Wenger the main implication of his analysis is for organizations. For the present authors, the design goals are pedagogical. The combination of pedagogical and organizational design principles offers the potential of innovative and potentially powerful applications.

Wenger's account of the stages of coming to belong to a community of practice parallels the stages of learning in our own account. His imagination stage, where the individual begins to identify the boundaries of a community of practice, involves a process of conceptualization. Equally engagement and construction are both about 'doing and discovering', and alignment and identification are stages which adjust understanding to a wider context.

A learning relationship exists when we learn from, or through, others. Such relationships will vary according to the characteristics of the groups involved, the context within which they operate, and the strength of the relationships. These relationships may be one of three different forms: one-to-one (e.g. parent to child); one-to-many (e.g. teacher to learners); and many-to-many (e.g. learning in peer groups or networks). The strength and effectiveness of the learning will also vary within the different kinds of relationships. These types of relationship vary according to the nature of the learning experience and are referred to as explorative, formative and comparative learning relationships, and they are described in more detail below.

An *explorative* learning relationship is about discovery. In our previous account it involves the discovery of other peoples' concepts, and for Wenger, the discovery of the practice boundaries. This relationship is often very descriptive (discovering the 'what' more than the 'why'), although some level of explanation will be involved, and asymmetrical in that most of the information flow is one way (from the 'outside in'). A *formative* learning relationship focuses more on the building of understanding through guided activity, and this is achieved through a constructive approach – the building and testing of hypotheses about the world, and about the nature of practice. This relationship is likely to be balanced or symmetrical, with outputs from the learner matched by feedback on performance. The third kind of learning relationship, *comparative*, characterizes a relationship that occurs once a level of expertize has been achieved or when an individual becomes accepted as more than a peripheral member of a community of practice. It is a relationship that allows the learner to *identify* their state of knowledge with others, or to *align* their practice with that of a community or organization. The primary purpose of a comparative relationship is not necessarily to acquire new knowledge, but to position and adjust existing knowledge by comparison with other knowledge states. The asymmetry is therefore in the opposite direction to the explorative stage (i.e. 'inside out'), and is less about acquisition and more about tuning and maintenance of the knowledge or practice. Such relationships may take on a defensive nature that can account for both the emotional charge of some of the dialogue that takes place and the resistance to change of some communities of practice.

Social network theory provides us with a method for describing and modelling learning relationships within given boundaries. The nodes in such a network represent the learners, and the links between them indicate that a relationship exists. Further, the concept of learning relationships recognizes that such relationships in themselves exist within a wider network. A relationship can be defined in terms of the *distance* between the two nodes. Learning relationships (regardless of type) are all first-order relationships. Clearly, second-order and other more distant relationships will influence these relationships and their requirements. For example, the relationship between a student and a teacher is, in terms of learning, a first-order relationship. However, that relationship will be influenced by the teacher's relationship with the year co-ordinator (second-order), and the year co-ordinator's relationship with the headteacher (third-order) and so on. It is assumed that the effective boundary of a learning relationship does not exceed a third-order relationship, although a community of practice could extend beyond that (see Figure 1). The community formed by a bounded learning relationship can be regarded as a micro-community of learners, within a community of practice. As well as determining the boundaries of the micro-community, another reason for measuring the different orders of relationship is that it may help to identify ways of supporting a first-order relationship. The basic rationale for a social network analysis is to focus attention on the relationships that exist, and to identify those that do not.

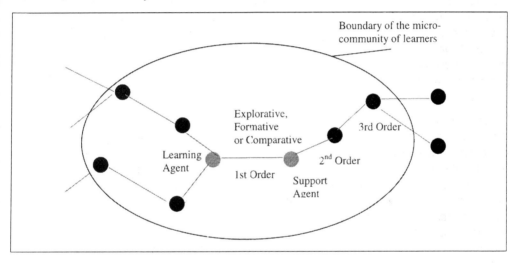

Figure 1: A typical representation of a learning relationship

These characteristics of the learning relationship are summarized below:

- Forms: one-to-one; one-to-many; many-to-many;
- Types: explorative, formative, comparative;
- Distance: first-order; second-order; third-order;
- Context: social groups; socio-political climate; wider community of practice.

The nodes within a learning relationship or the wider learning network are referred to as agents. The learner in the relationship is called a 'learning agent', and the facilitator is called a 'support agent'. It is important to distinguish between the different functions provided by the learning agent. Any agent must be a knowledge resource, able to sustain a dialogue with the learner, and able to guide and support the learning tasks. These are the characteristics which have previously been used to specify the requirements for an intelligent tutoring system (Wenger, 1987). Any agent must be able potentially to satisfy these criteria, although in any particular learning relationship these conditions may not be met. A crucial issue to address is the extent to which the characteristics of an agent in a learning relationship can be provided through technology. This is a fundamental question about the role of learning technology, under which most of the design issues for computer-based learning environments can be subsumed. What aspects of effective learning support absolutely require a human relationship?

What next?

Our next steps are to operationalize the concept and to collect empirical evidence to test, refine or refute it. We intend to explore the learning relationships of sixteen-year-olds attending schools in three different European countries (UK, Finland and Portugal). By using countries from the north, middle and south of Europe we are also hoping to identify any key cultural variables that may influence the nature and form of the learning relationships.

Learning relationships will be identified using a questionnaire, and represented visually in a learning network using a proprietary drawing tool. The drawing tool also supports annotations of the relationships with both specific data (e.g. descriptions of the learners) out to more general data (e.g. descriptions of the schools and cultures). Once the main support agent of the learner has been identified, then the relationship between the learner and support agent (e.g. teacher, parent or friend) can be classified. The two agents undertake a task and their learning behaviours are classified as one of the three learning types through the use of a checklist. The relationship type is then noted on the learning relationship diagram. Data are also being collected about the individual learner, their families, school and wider communities.

Once all the data have been collected these will be analysed using a multi-level modelling technique (Goldstein, 1995). This technique allows one to assess the different factors affecting the relationships. For example, strong learning relationships may only be found for learners from certain types of families or who attend schools with a particular learning ethos.

If learning relationships prove to be a powerful factor in learning, then the concept will be used to rethink how we define online educational services. Future services that are sensitive to learning relationships may look for ways to optimize configurations so that key agents are connected and their relationship is maximized to enhance learning. It is planned to build and test such a prototype service for home-school communities in the UK, Finland and Portugal.

The emphasis of this empirical work is to explore whether or not learning relationships can provide us with a sufficiently robust analytical model for determining requirements for, and evaluation of, learning networks. Where such networks exist, the concept can help us identify relationships that should be formed, optimized or even terminated. An IP-based service based on learning relationships may be just what is required to move us away from the overemphasis on content or the 'teacher-in-the-box' services that are currently prevalent in the market.

Conclusions

By subjecting the framework developed in our earlier work (Mayes and Fowler, 1999) to the perspective offered by Wenger's work on communities of learning we have shifted the focus of our work in the following ways:

- moving the emphasis of learning away from the 'what' we learn to the 'who' we learn from;

- more directly addressing the key factor of motivation to learn;

- providing a bridging framework from which both pedagogical and organization design can be derived;

- offering insights into how best to manage change within our learning organizations.

This new emphasis helps to reinforce the point that learning within organizations and within traditional educational settings requires a common explanatory framework.

It remains to be seen whether we can move from the conceptual framework described in this paper to the specification of a design methodology which properly embeds the technological support of learning in the wider social context in which the identities of learners are shaped.

Acknowledgements

This project is partially funded by the European Institute for Research and Strategic Studies in Telecommunication (EURESCOM) and involves partners from BT (UK), Portugal Telecoms, Telenor (Norway); Finnet Group (Finland), Slovakia Telecoms and Icelandic Telecoms. The authors are also grateful to the many colleagues who have contributed through discussion. In particular, we like to thank the members of the Vicarious Learning Project based in Glasgow Caledonian and Edinburgh Universities and BT's Education and Training Research team at Adastral Park. We are also grateful to Christina Knussen for helpful advice on Social Identification literature.

Notes

[1] Chris Fowler also works at the Institute of Education, University of London, in a half-time capacity.

[2] The original framework was based on a simple goal-action feedback loop called the conceptualization cycle. As with Miller, Galanter and Pribram's TOTE units (1960) it was seen as a building block from which more and more complex descriptions of learning could be progressively constructed. (Larger TOTE units could be built from smaller, a process that modelled the way in which practice builds larger skill sequences.) The framework referred to here is the second-order description, where the action stage was unpacked into stages of conceptualization, construction and dialogue. The framework is intended less as a description of learning and more as an illustration of design principles that can be distilled from a range of psychological, social and educational literatures.

References

Anderson, J. R., Boyle, C. F., Farrell, R. and Reiser, B. J. (1987), 'Cognitive principles in the design of computer tutors', in P. Morris (ed.), *Modeling Cognition*, New York: Wiley.

Barab, S. and Duffy, T. (in press), 'From practice fields to communities of practice', in D. Jonassen and S. Land (eds.), *Theoretical Foundations of Learning Environments*, Mahwah, NJ: Erlbaum.

Brown, J. S., Collins, A. and Duguid, P. (1989), 'Situated cognition and the culture of learning', *Educational Researcher*, 18, 32–42.

Brown, S. (1999), 'Virtual university: real challenges', in B. Collis and R. Oliver (eds.), *Proceedings of ED-MEDIA 1999: World Conference on Educational Multimedia and Telecommunications*, AACE.

Cognition and Technology Group at Vanderbilt (1990), 'Anchored instruction and its relation to situated learning', *Educational Researcher*, 19, 2–10.

Cole, M. and Scribner, S. (1974), *Culture and Thought: A Psychological Introduction*, New York: Wiley.

Collins, A., Brown, J. S. and Newman, S. E. (1989), 'Cognitive apprenticeship: teaching the crafts of reading, writing and mathematics', in L. B. Resnick (ed.), *Knowing, Learning and Instruction: Essays in Honour of Robert Glaser*, Hillsdale, NJ: Erlbaum.

Fowler, C. J. H. and Mayes, J. T. (1997), 'Applying telepresence to education', in *BT Technology Journal,* 14, 188–95.

Garton, L., Haythornthwaite, C., and Wellman, B. (1997), 'Studying on-line social networks', *Journal of Computer Mediated Communication*, 3 (1).

Glaser, R. (1990), 'The re-emergence of learning theory within instructional research', *American Psychologist*, 45, 1, 29–39.

Goldstein, H. (1995), *Multilevel Statistical Models*, London: Edward Arnold.

Hog, M. A. and Adams, D. (1988), *Social Identification*, London: Routledge.

Kaye, A. R. (ed.) (1992), *Collaborative Learning through Computer Conferencing: The Najaden Papers*, Heidelberg, FRG: Springer Verlag.

Laumann, E., Marsden, P. and Prensky, D. (1983), 'The boundary specification problem in network analysis', in R. Burt and M. Minor (eds.), *Applied Network Analysis*, Beverly Hills, CA: Sage.

Laurillard, D. (1993), *Rethinking University Teaching: A Framework for the Effective Use of Educational Technology*, London: Routledge.

Lave, J. and Wenger, E. (1991), *Situated Learning: Legitimate Peripheral Participation*, Cambridge: Cambridge University Press.

Lave, J. (1988), *Cognition in Practice*, Cambridge: Cambridge University Press.

Lave, J. (1997), 'The culture of acquisition and the practice of understanding', in

Kirshner, D. and Whitson, J. A. (eds.), *Situated Cognition: Social, Semiotic and Psychological Perspectives*, Mahwah, NJ: Erlbaum.

Mayes, J. T. and Fowler, C. J. H. (1999), 'Learning technology and usability: a framework for understanding courseware', *Interacting with Computers*, 11, 485–97.

Mayes, J. T. (1995), 'Learning technologies and Groundhog Day', in Strang, W., Simpson, V. B. and Slater, D. (eds.), *Hypermedia at Work: Practice and Theory in Higher Education*, Canterbury: University of Kent Press.

McKendree, J., Stenning, K., Mayes, J. T., Lee, J. and Cox, R. (1998), 'Why observing a dialogue may benefit learning', *Journal of Computer Assisted Learning*, 14, 110–19.

Miller, G., Galanter, E. and Pribram, K. (1960), *Plans and the Structure of Behavior*, New York: Holt, Reinhart & Winston.

Papert, S. (1990), 'An introduction to the fifth anniversary collection', in Harel, I. (ed.), *Constructionist Learning*, MIT Media Laboratory, Cambridge: MA.

Philips, V. and Yager, C. (1998), *Best Distance Learning Graduate Schools: Earning your Degree without Leaving Home*, Princeton Review: Random House.

Resnick, L. B. (1987), 'Learning in school and out', *Educational Researcher*, 16, 13–20.

Savery, J., and Duffy, T. (1996), 'Problem-based learning: an instructional model and its constructivist framework', in Wilson, B. (ed.), *Constructivist Learning Environments: Case Studies in Instructional Design*, Englewood Cliffs, NJ: Educational Technology Publications.

Senge, P. (1994), *The Fifth Discipline Fieldbook: Strategies and Tools for Building a Learning Organisation*, New York: Doubleday.

Sfard, A. (1998), 'On two metaphors for learning and the dangers of choosing just one', *Educational Researcher*, 27, 4–13.

Suchman, L. (1987), *Plans and Situated Actions: The Problem of Human-Machine Interaction*, Cambridge: Cambridge University Press.

Turner, J. C. (1991), *Social Influence*, Milton Keynes: Open University Press.

Wellman, B. and Berkowitz, S. D. (1988), *Social Structures: A Network Approach*, Cambridge: Cambridge University Press.

Wenger, E. (1987), *Artificial Intelligence and Tutoring Systems: Computational and Cognitive Approaches to the Communication of Knowledge*, Los Altos, CA: Morgan Kaufmann.

Wenger, E. (1998), *Communities of Practice: Learning, Meaning and Identity*, Cambridge: Cambridge University Press.

Update – Learning relationships from theory to design

If learning relationships are to underpin design, we must ask ourselves if the construct is valid. Typically, in order to determine the validity of a theoretical concept, hypotheses are generated and tested, or a more explorative strategy is adopted. Our approach involves a combination of the two. So far, we have tested the general framework on groups of 16-year-olds in the UK, Finland and Portugal.

The first stage of the empirical work involved identifying the children's learning agents. The children were asked, after reading some educational scenarios, whom they would choose to help them solve the problems posed, and first-, second- and third-order support agents were identified. The second stage involved classifying the learning activity taking place in the relationship between learner and first-order support agent. Behaviours were captured over a 40-minute learning task, and the learning classified as explorative, formative or comparative. Finally, contextual information about the children, their parents, attitudes to school and friendship patterns was collected. These data were then modelled using a multi-level modelling technique developed by Goldstein (1995).

In general, enough evidence was found to support the theoretical notion of learning relationships, although the data imply that our initial framework was too simplistic. Important aspects of the behaviour of the children are sensitive to cultural differences, and there were some markedly different patterns associated with the Portuguese group. Overall, however, the results confirmed that changes occur in the nature of the learning as the learners become more expert, and that the influence of a particular learning agent (teacher, father, mother, peer) indeed depends on the kind of learning dominant at a particular stage. Nevertheless, there is a complex interplay between cultural and contextual factors that directly influences the interpretation of the characteristics of a learning relationship. The support roles played by parents, teachers and peers are themselves highly influenced by social and cultural variables.

We should now perhaps pause to reconsider where this research should lead. At one level, our work is aimed at demonstrating that by focusing on aspects of educational settings traditionally underplayed in design, we are more likely to build effective support systems. Yet this is of little value unless it can produce principles on which to base design decisions. Our theoretical framework must capture enough of the complexity to support specific guidelines, but also be sufficiently robust to underpin high-level design decisions.

Reference

Goldstein, H. (1995), *Multilevel Statistical Models,* London: Edward Arnold.

Contact author

Chris Fowler is a cognitive psychologist. He joined BT's Research Laboratories in 1990; before that he worked in various Higher Education Institutes. Since 1998, he has worked in a half-time capacity as manager of BT's Education and Training research programme, and in the other half of his time, as Director of the Future Learning Centre, Institute of Education, University of London. [*chris.fowler@bt.com*]

The impact of educational technology: a radical reappraisal of research methods

P. David Mitchell
Graduate Programme in Educational Technology, Concordia University, Montreal, Canada

Initially published in 1997

How can we decide whether some new tool or approach is valuable? Do published results of empirical research help? This paper challenges strongly entrenched beliefs and practices in educational research and evaluation. It urges practitioners and researchers to question both results and underlying paradigms. Much published research about education and the impact of technology is pseudo-scientific; it draws unwarranted conclusions based on conceptual blunders, inadequate design, so-called measuring instruments that do not measure, and/or use of inappropriate statistical tests. An unacceptably high portion of empirical papers makes at least two of these errors, thus invalidating the reported conclusions.

Introduction

The practical problem which motivates this paper is that of deciding – on the basis of published research – whether to adopt some new device, procedure or paradigm thought likely to improve education. What models, methods or media are likely to be most useful? From the invention of the printing press to multimedia software, educators have adopted unproven aids and fads. Researchers usually claim each new device or procedure to be at least as effective as its predecessor. How valid is all this research? How to decide? A typical view is: *Design an experiment to observe the effects of your treatment. Any book on research design and statistics will show you how.* But will it? What essential aspects of educational measurement and research must we consider?

Measurement or sorcery?

Suppose we wish to conduct research on media-based learning as a function of different learning styles. Let us assume that we decided operationally to define relevant styles with a commonly used questionnaire (for a more complete discussion of learning styles and problems of identifying them, see Mitchell, 1994). A typical questionnaire asks a series of questions that one answers on a scale of possible responses ranging from, for example, *strongly agree* to *strongly disagree*. But let us examine a measurement issue first.

Measurement: neglected rules

From mathematics, the theory of numbers and the theory of measurement provide the foundation upon which educational measurement and statistics must rest if the latter are to be more than superficial and deceptive. The axiom of identity requires that: each question be equivalent to each of the others; two people with the same score must have comparable abilities; and equal differences between scores be equivalent. The result, if the assumption of equivalence is not violated, is similar to a thermometer; a one-degree difference is the same unit regardless of the starting temperature. Such equal interval scales (see Stevens, 1946) are common in science but not in education.

Instruments presumed to measure some variable like an attitude, opinion or even knowledge, seldom have equivalent questions. Moreover such technical refinements as reliability, validity or internal consistency fail to satisfy this axiom of identity. There is no guarantee that identical scores represent students with identical answers; indeed, it is very unlikely. Yet researchers usually treat questionnaires dealing with linguistic concepts (for example, comprehension or learning style) as if they were sharply defined interval scales. Most scales used in educational research actually are ordinal scales and therefore do not meet the mathematical preconditions for the statistical manipulations commonly used (Liebetrau, 1983).

Consider the questions on commonly used 'instruments' purported to measure learning styles (there are over 100, but see Entwistle, 1981). Typically in such scales, response categories are ranked in order of importance to the researcher. Ranking may begin with *definitely disagree* assigned a rank of 1, and so on to 5. The test creator usually considers this rank to be a 'score' for each question so that he or she can perform statistical analyses on the numbers. Another conceptual blunder is to add the so-called scores for several questions to get a 'total' for that subscale which carries a label supposedly denoting a variable (for example, a particular learning style). This occurs despite the items' appearing to violate the axiom of identity; thus they cannot be added even if each scale were interval.

With a wave of a magic wand (accompanied by the incantation, 'let us assume . . .') it seems that we can represent a statement ('I disagree that . . .') by a number which is not simply a symbol or identifier of a position in a sequence but a quantity. But can we? Is it justified? Mathematically, the difference between 4 and 3 is equivalent to 2 minus 1 or 3 minus 2, but is it correct to say that the difference between my saying that 'I definitely agree . . .' and 'I agree with reservations . . .' is the same as between 'impossible to give a definite answer' and 'I disagree with reservations . . .'? Logically, all we can assume with a ranking of categories is the sequence. Statistics deals with numbers, not what they represent. If numbers, as collected and assigned, violate epistemological or mathematical requirements, the analysis will produce mathematically correct results. But what do they mean?

Measurement or intellectual pollution?

Many published 'measurement instruments' were generated by factor analysis. Surely this procedure justifies the scale and its scoring? Space limitations permit no discussion here, but this argument should be interpreted in the light of Patrick Meredith's pithy comments about Spearman's contribution to the topic:

> What is disturbing is that Spearman's 'factorial' concept, whose epistemological basis is riddled with fallacies, not only took off but came to dominate the psychological and educational skies [. . .] Instructional Science has a decontamination job on its hands, to disperse the intellectual pollution created by a whole profession reared on a contempt for real information and a superstitious worship of false quantification. (Meredith, 1972, p. 16)

Is it possible that educational researchers have a 'contempt for real information and a superstitious worship of false quantification'?

Information, numbers and statistics

In contrast to Stevens (1946) and his followers, I assert that measurement is not just the assignment of numbers to things according to specified operations. The purpose of measurement is to reduce the variety of some part of reality which we observe, whether directly or through some information-gathering activity, to yield summarizing information that is accurate, precise and general. Usually our intention is to answer a question or to support a decision.

Pseudo-measurement

What too frequently happens is that the 'score' produced by the 'scoring key' (by illegitimately summing ranks of ordinal measures) is treated as if it were quantitative information about that variable for each person. Textbooks and professors often claim that it is all right to treat Ordinal Scales as if they were Interval Scales because their test of significance is so robust that it is unlikely to lead to improper conclusions. How credible is this? Note that 'robust' is contextual, not fixed, contrary to a common myth. And any violation of a test's prerequisites alters its probabilities of Type I and II errors. Moreover, the powers of some parametric tests have been shown to diminish to zero under violations of mathematical assumptions of the test (Bradley, 1982).

If we play games with epistemological underpinnings and mathematical prerequisites, the consequences are unknown, and our analysis could be meaningless. Lakatos, a philosopher of science, dismissed our typical use of statistical techniques to produce 'phoney corroborations and thereby a semblance of "scientific progress" where, in fact, there is nothing but an increase in pseudo-intellectual garbage' (Lakatos, 1978, p. 88).

Insignificance of statistical significance

Consider this quotation from a typical textbook:

> Tests of statistical significance are used to help researchers to draw conclusions about the validity of a knowledge claim. [. . .] If the null hypothesis is rejected, we conclude that the knowledge claim (i.e. the research hypothesis) is true. If the null hypothesis is accepted, we conclude that the knowledge claim is false. (Meehl, 1978, p. 622)

We usually are told to reject the null hypothesis if the difference is 'significant' (i.e. $p < 0.05$). But consider Meehl's summarizing statement: '[. . .] if you have enough cases and your measures are not totally reliable, the null hypothesis will always be falsified, regardless of the truth of the substantive theory'. Despite its prevalence, null-hypothesis significance testing has been criticized for half a century. Lykken's conclusion can guide us:

> Finding of statistical significance is perhaps the least important attribute of a good experiment; it is never a sufficient condition for concluding that a theory has been corroborated, that a useful empirical fact has been established with reasonable confidence – or that an experimental report ought to be published. The value of any research can be determined, not from the statistical results, but only by skilled, subjective evaluation of the coherence and reasonableness of the theory, the degree of experimental control employed, the sophistication of the measuring techniques, the scientific or practical importance of the phenomena studied, and so on. (Lykken, 1968, p. 159)

Inappropriate use of parametric statistics

The majority of reported studies reflect the parametric statistical techniques taught in educational research courses for decades. These 'assume' (i.e. have as a mathematical pre-requisite) a normal distribution of the data in a population, even though educational researchers seldom know much about the population beyond the small sample taken in their experiment. (Paradoxically, the more experimental control and the greater the treatment effect, the less approximately normal the distribution will be, and the test's power is attenuated.)

In parametric statistical analysis, what we find is an answer to this question: 'IF the null hypothesis (of no difference between groups) is true, AND if both normality and homogeneity of variance are true, what is the probability that we would find a difference at least as large as that which we observed, if we had randomly drawn two samples from that population?' This is easy to compute, but it is only correct if the prerequisite assumptions are true. Alas, a major problem exists: we do not know if the null hypothesis is true, if the data is normally distributed, if homogeneity of variance is true, or even if the data is random (though our design may reassure as about randomness). So what happens? The researcher can only act as if this were the case by assuming it to be so. Much of the time, it is not.

Simple observation of published means and standard deviations reveals many which clearly cannot be normal distributions (for example, relatively large standard deviations, often nearly as large as, and sometimes several times larger than, the mean; non-symmetrical shape). Lohr *et al* (1995) reported several differences in their 'comprehensive evaluation' of a hypertext model for teaching, but their conclusions came from Analyses of Variance, some of which were 'significant'. Although Meehl (1978) shows that one should not even make such claims with multiple tests, let us examine the comparisons. The authors offer 21 means and standard deviations. Given that a normal distribution is symmetrical and extends three standard deviations above and below the mean, it is noteworthy that only one of the 21 groups could be approximately normal. Eight involved standard deviations as large as or larger than the mean. Clearly these do not describe normally distributed data, and a parametric test should not be used. Unlike those authors, we can conclude nothing except that the analysis was inappropriate. Over half the empirical papers in recent issues of several journals make the same mistake.

As Krauth showed:

> Examination of real data reveals that the assumption of a normal distribution is not justified in the majority of cases. Empirical distributions are seldom symmetrical, a necessary assumption for normality. Furthermore, empirical distributions tend to have heavier tails [. . .] One argument often used to justify [parametric tests] is that these tests are quite robust [. . .] [an argument] based on some old studies [about which problems exist]. (Krauth, 1988, p. 15)

Bradley concurs:

> A fantastic folk-lore sprang up among research workers: [. . .] distribution-free tests were regarded as second-class statistics, hopelessly inferior to parametric statistics [whether or not they meet their assumptions]. The efficiency of distribution-free relative to classical tests has been investigated under common non-parametric conditions, i.e. non-normal populations and/or heterogeneous variances, and the new statistics have often proven superior, sometimes infinitely so. (Bradley, 1982, p.13)

Note, too, that if you choose to use null hypotheses, those associated with distribution-free statistics are more general and thus more realistic than those of classical statistics. So if you wonder: 'When should I use distribution-free statistical methods?', the answer is: whenever possible.

Where do we go from here?

It seems likely that few of the published papers in our field meet all or even most of the requirements of scientific research (theory-oriented, conceptually clear, measurements consistent with number theory and measurement theory, preconditions for parametric statistics met or distribution-free statistics used, appropriate logic, replicative validity). As support for decisions or research, most published results and interpretations probably should be discarded. This may be too extreme but it seems to be a good starting-point.

Educational technology may be riding a wave – the wave of pseudo-science. This wave moves out of universities and through our journals. It is generated by the use of textbooks that perpetuate an unthinking, cookbook approach to scientistic rather than scientific inquiry. A radical shift in our own thinking is essential.

The challenge facing us is complex. I offer a few ideas merely as conversation starters. It strikes me that the academic preparation of educational researchers should be reformed – radically. I suggest that the research student be expected to master cybernetic principles, systems modelling, probability theory, epistemology and the philosophy of science before embarking on the study of research design and analysis. And the first stage in educational research should focus on epistemological, not statistical or even design, issues. Indeed, the first statistics course should focus on probability and non-parametric statistics with parametric statistics left for later. As for non-researchers, instead of wasting time studying cookbook-based parametric statistics, they might spend their time better studying the logic of scientific inference and the fundamentals of modelling and measurement so that they may be able to detect and reject pseudo- and quasi-pseudo-scientific research when they encounter it in our periodicals. And if they study statistics, let them begin with non-parametric statistics which is so much easier to grasp and more generally applicable.

But this is not sufficient. Journal editors and their referees need to modify their tendency to accept any quantitative article that contains a statistically significant outcome and focus more on the quality of thinking, the validity of measurement and appropriateness of qualitative and quantitative methods. Even better, they should encourage replications. Conceivably, a comparison of names of authors of pseudo-scientific articles with those on the editorial boards of our journals might reveal another problem, and journal editors may have to find new referees.

Conclusion

Much of our cherished research is pseudo-scientific with unwarranted conclusions. In considering the standard approach it is instructive to bear in mind Stafford Beer's comment:

> A paradigm is a model that exhibits a closed logic and thus resists change [. . .] To create change, you must challenge not only the models of unreality but the paradigms that underwrite them. Dangerous work. (Beer, 1988)

The paradigm informing most empirical educational research is an input-process-output model which assumes a one-way direction of causality from independent to dependent variables. This questionable model lies at the heart of t-tests, correlations, analysis of variance, multiple regression, discriminant function analysis, etc. In many papers, the mathematical prerequisites of these statistical tests are not met, thus invalidating the results.

Another fundamental flaw involves passing subjective judgements off as measurements (by attaching numbers to constructs for which none of the properties of measurable magnitude are met), then combining these, violating stringent mathematical requirements and producing meaningless results. Finally, treating linguistic variables (for example, computer-aided learning, hypertext, multimedia, learner control) as if they are precise and comparable may confuse rather than inform.

I have tried to show that our research needs to begin with conceptual issues but that our biggest problem is the way in which research is carried out. I have tried to stimulate thinking about and questioning the models and underlying paradigms that permeate our exciting field of study and practice. Whether or not you ever carry out a research programme, you are certain to encounter claims (in the mass media as well as in textbooks and journals) that are not justified or justifiable. Now you can reject them.

References

Beer, S. (1988), *Address to Convocation*, Concordia University, Montreal.

Bradley, J. V. (1982), *Distribution-Free Statistical Tests*, Englewood Cliffs, NJ: Prentice-Hall.

Entwistle, N. J. (1981), *Styles of Learning and Teaching*, Chichester: John Wiley.

Krauth, J. (1988), *Distribution-Free Statistics*, Amsterdam: Elsevier.

Lakatos, I. (1978), 'Falsification and the methodology of scientific research programmes' in Worrall, J. and Currie, G. (eds.), *The Methodology of Scientific Research Programs*, Philosophical Papers, vol. 1: Imre Lakatos, Cambridge: CUP.

Liebetrau, A. M. (1983), *Measures of Association*, London: Sage.

Lohr, L., Ross, S. M. and Morrison, G. R. (1995), 'Using a hypertext environment for teaching process writing: an evaluation study of three student groups', *Educational Technology Research and Development*, 43 (2), 33–51.

Lykken, D. T. (1968), 'Statistical significance in psychological research', *Psychological Bulletin*, 70, 151–9.

Meehl, P. (1978), 'Theoretical risks and tabular asterisks: Sir Karl, Sir Ronald, and the slow progress of soft psychology', *Journal of Consulting and Clinical Psychology*, 46 (4), 822.

Meredith, P. (1972), 'The origins and aims of epistemics', *Instructional Science*, 1 (1), 16.

Mitchell, P. D. (1994), 'Learning style: a critical analysis of the concept and its assessment', in Hoey, R. (ed.), *Aspects of Educational Technology XXVII*, London: Kogan Page.

Stevens, S. S. (1946), 'On the theory of scales of measurement', *Science*, 103, 677–80.

Update – The impact of educational technology: a radical reappraisal of research methods

There is a story that Galileo's contemporaries refused to look through his telescope to see sunspots because they did not believe there were spots on the sun. Similarly, educational researchers ignore evidence spanning six decades that delineates abuses of measurement theory and statistical analysis. A cursory survey of recent journal articles reveals little change (e.g. parametric analyses despite standard deviations larger than the mean; principal component analysis of Likert items with n=19).

Those who engage in so-called quantitative educational research, including research on information and communications technology thought to be important for learning, generally ignore the issues outlined in my paper reproduced in this volume. Texts on educational research methodology and statistics seldom address the fundamental ideas. When they do, it is usually a matter of relegation to a short comment, or even a dismissal, rather than a genuine discussion.

Even awareness is no guarantee that the researcher understands the fundamental issues. Consider this note from an educational technology Faculty member (who shall remain anonymous): 'With regard to the parametric non-parametric issue, I think [you are] technically correct but given the glaring conceptualization weaknesses of many research designs etc., that is a relatively minor issue. If effect size is properly determined, then null hypothesis testing probably does not do too much harm.' Of course, many research designs suffer from conceptual blunders, but if their publication is compounded by flaws in reasoning, measurement or misuse of parametric statistical tests, I submit that this is not a relatively minor issue, either for the author or for the journal to which work is submitted. Nor is it minor for the growth of knowledge. Null hypothesis significance testing, like believing sunspots do not exist, may not harm anyone – provided appropriate statistical models are used – but it is fundamentally unsound and unscientific, and fails to tell us what we wish to know. If parametric tests are used inappropriately, the result may be (as Lakatos observed) an increase in pseudo-intellectual garbage. Conclusions drawn from much of our research must be scrutinized carefully and replicated before we accept them.

Educational journals should adapt the Statistical Guidelines for Contributors to Medical Journals (Gardner and Altman, 1989), but their checklist question – 'Were the statistical analyses used appropriate?' – should be expanded.

Reference

Gardner, M. J. and Altman, D. G. (1989), *Statistics with Confidence*, London: British Medical Journal.

Contact author

P. David Mitchell's intellectual background is extensive, with courses and 118 publications in several fields, but he thinks of himself primarily as a psychologist turned cybernetician/ operational researcher focusing on education. He retired early as a Professor of Educational Technology at Concordia University, Montreal, and now is an independent researcher, writer and consultant. [*mitchel@alcor.concordia.ca*]

CAL evaluation: future directions

Cathy Gunn
Centre for Professional Development, University of Auckland, New Zealand

Initially published in 1997

Formal, experimental methods have proved increasingly difficult to implement, and lack the capacity to generate detailed results when evaluating the impact of CAL on teaching and learning. The rigid nature of experimental design restricts the scope of investigations and the conditions in which studies can be conducted. It has also consistently failed to account for all influences on learning. In innovative CAL environments, practical and theoretical development depends on the ability fully to investigate the wide range of such influences. Over the past five years, a customizable evaluation framework has been developed specifically for CAL research. The conceptual approach is defined as Situated Evaluation of CAL (SECAL), and the primary focus is on quality of learning outcomes. Two important principles underpin this development. First, the widely accepted need to evaluate in authentic contexts includes examination of the combined effects of CAL with other resources and influential aspects of the learning environment. Secondly, evaluation design is based on a critical approach and qualitative, case-based research. Positive outcomes from applications of SECAL include the easy satisfaction of practical and situation-specific requirements and the relatively low cost of evaluation studies. Although there is little scope to produce generalizable results in the short term, the difficulty of doing so in experimental studies suggests that this objective is difficult to achieve in educational research. A more realistic, longer-term aim is the development of grounded theory based on common findings from individual cases.

Experiments that failed

Scientific, experimental methodology was previously considered to be the only acceptable approach to educational research. Two important principles of experimental design are:

- to balance individual differences within study populations and so achieve generalizable results,

and

- to attempt to isolate the effects of a single resource for evaluation purposes.

Problems with this approach were reported in the literature of the 1970s (Elton and Laurillard, 1979; MacDonald and Jenkins, 1979) when the influence on learning of individual and contextual factors was recognized. Similar issues emerged during the 1980s and early 1990s, (Bates, 1981; Spencer, 1991) when the inability to identify which single or combined factors supported learning became a recurrent problem. It was clear that prior knowledge, approaches to learning, provision of appropriate scaffolding, complementary combinations of resources and various contextual factors all influenced the quality of learning outcomes. It was concluded that evaluations must be designed to account for these factors, rather than to balance or disregard them as was previously the norm (Kemmis, 1987, Gunn, 1995).

Another problem stemmed from the belief that single studies involving large sample sizes were necessary to produce meaningful results. The rather indiscriminate choice of study populations required to produce the requisite numbers frequently resulted in low motivation and levels of perceived relevance of evaluation tasks to personal and educational goals (Draper *et al*, 1996; Gunn, 1996). This suggested that the true potential for learning with CAL could not be reliably assessed unless its use formed an integral part of a course, and evaluations involved only the students on that course. It was thus concluded that the more specific aspects of CAL evaluation could not be served by a general and inflexible research methodology originally designed to measure the uniform and largely predictable behaviour of organisms in the physical sciences.

The basis of an alternative methodology

In the context of the work reported here, development of a suitable methodology began with a review of educational research literature. Critical theory (Carr and Kemmis, 1986), critical ethnography (Angus, 1986) and qualitative methodology (Denzin and Lincoln, 1994) were adopted as the basis for grounded development of the Situated Evaluation of CAL (SECAL) framework. The stages of development are described elsewhere (Gunn, 1995; 1996). Within the situated approach, the standard range of objective and subjective research methods are used, as appropriate, for evaluation study design, i.e. observation, field notes, log data, interview, questionnaires, outcomes analysis, results comparisons, attitude surveys, expert review, and discussion groups.

The SECAL approach is opportunistic in recognizing that situation-specific factors such as logistical and ethical constraints will determine what subset of qualitative methods is available, and of these methods, which are the most appropriate in a particular case. The case-study method described by Yin (1991) is the basis for study design, and from this comes the longer-term objective to develop grounded theory (Glaser and Strauss, 1967). Action research (Zuber-Skerrit, 1990) describes the preferred, collaborative approach involving all interested parties in CAL development and integration initiatives. It also defines the action-reflection-modification cycle that models the dynamic process of educational change.

The concept of situated evaluation

The concept of situation is described in relation to knowledge and learning by Brown *et al* (1989): 'Knowledge is situated being in part a product of the activity, context and culture in which it is developed and used.' Although there is some debate about the precise nature of situatedness, if the concept is accepted at all, evaluation should not assess whether CAL alone supports learning or enhances its effectiveness. It must examine the effects of particular resource combinations within specific learning contexts. It was never considered necessary to evaluate the impact of a book or lecture in isolation, so the justification for doing so with CAL is hard to see.

The range of potential influences on learning is broad, and includes factors which are both intrinsic and extrinsic to any particular study resource. To accommodate all possibilities, SECAL includes the concepts of *evaluation in context* and *evaluation of context*. *Evaluation in context* refers to study of the primary effects of using CAL programs with other resources and forms of support. The proper integration of CAL into courses is crucial. *Evaluation of context* examines factors related indirectly to a CAL program or the immediate learning environment, but ones which can still influence integration at an institutional level, and so impact on learning outcomes. Factors related to levels of institutional support for acquisition, development and use of CAL fall into this latter category. There is also a dynamic aspect of SECAL which supports recommendation and implementation of beneficial changes to learning environments.

Authenticity and context

The importance of contextual influences implies that authenticity in study design should be a non-negotiable factor. Evaluation is consequently limited in scope and frequency by the number of target users and available opportunities for the effects of CAL to be evaluated as a fully integrated part of a course. Although these limitations may appear to be rather restrictive, the comprehensive and theoretically supported SECAL framework has been successfully applied in very different circumstances and has produced relevant and meaningful results. The experimental objectives of generalizable results and theory generation are not ruled out – they just take longer to achieve. In view of the strict relevance of outcomes, the compromise is worthwhile.

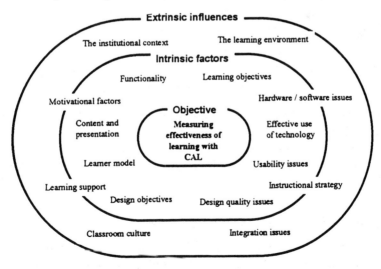

Figure 1: SECAL – a framework for situated evaluation of CAL

SECAL element	Case-specific details
Statement of evaluation objectives	To assess the impact on learning of using CAL to replace some practical laboratory work: • examine the case for investment in more workstations and software licences; • explore the potential for development of stand-alone CAL tutorial exercises; • identify areas for further investigation through qualitative, quantitative or longitudinal studies.
Learning objectives	Students should gain a basis for understanding the biomechanics of movement and the nature and avoidance of sports injuries. The evaluation tasks measured learning by the ability to locate, identify and describe interconnections of various structures.
Hardware/software issues	The product was bought in and so offered no scope for modification. The hardware specification complied with what was already available within the department. However, licence limitations required it to run across a network, and speed, functionality, etc. had not been tested under these conditions.
Effective use of technology	The program presents clear and logically constructed representations of the systems, structures and layers of the human body. Views can be rotated, expanded and dissected in layers. A fair impression of a three-dimensional structure is given by something that is in fact two-dimensional. Access to written and spoken versions of the entire, complex terminology is provided.
Design objectives	These were not measured because the product was bought in. The developers' stated objectives were a useful product for all aspects of medical education.
Design quality, functionality and other usability issues	The program was of accepted high quality, scored well on usability factors, successfully engaged interest, and allowed tasks to be completed in a logical manner.
Instructional strategy	There was no built-in instructional strategy so this was provided by the context of use, i.e. • tailoring the work to an appropriate level of difficulty; • providing 'scaffolding' for novice computer users and those less confident with the subject; • promoting peer support and the benefits of collaborative learning; • using a constructivist approach to learning with CAL.
Learner model	The general learner model was defined by the product developers. It was further specified by contextual factors.
Content and presentation	The content of the program was comprehensive, and presentation was of good, logical standard. Presentation of the program itself was defined by situational factors.
Learning support	Some support was provided by features of the program. Additional sources were from situational factors such as the task, situation within the course, lecture notes, diagrams and the presence of the lecturer.
Motivational factors	• Making learning goals explicitly relevant to the whole course and to assessment requirements; • the attractive appearance of the program; • presenting the task as a series of challenges and providing feedback.
Classroom culture	The group had little previous experience of independent learning, so gradual introduction in a small-group setting was favoured; attention was paid to design of a non-threatening situation for novice users.
Institutional context	There existed no clear institution strategy related to developments in CAL and support was inconsistent

Figure 2: The SECAL framework for evaluating a CAL program in a course on human anatomy

SECAL description

Figure 1 shows the elements included in the SECAL framework, the relative importance of each being determined by the evaluation objectives and the interests being served. It is a simple matter to customize the framework by weighting each element according to its relevance in a particular case.

Integration issues

The elements of the SECAL framework all depend to some extent on CAL being fully integrated into courses. There have been cases where well designed, educationally sound and accessible CAL fails to achieve the success its potential implies. This is often attributable to poor integration strategies, either at institutional or classroom level. Where CAL is simply one available option rather than a compulsory part of a course, take-up rates are frequently poor because there is no compelling reason for students to adopt the new study habits involved. Where staff are not committed to technological advances, little encouragement may be passed on to students. Equally, where institutions do not actively encourage staff to use innovative methods little incentive or support may be available to those who wish to do so. At a broader contextual level, there may be social, political and economic pressures which shape institutional policy on matters of technological change in a positive or negative way. Such influences may seem a long way from the 1990s classroom where students are required to use new technologies as aids to learning and communication, but the measurable effects clearly can extend across this entire range.

Applying SECAL

The scenario in Figure 2 presents an example of the SECAL framework applied to the evaluation of a newly introduced CAL program in a course on human anatomy. The data-collection methods included independent observation, field notes, analysis of task performance, expert opinion and group discussions. A full case study report has been published by Gunn (1996).

Conclusions

The very brief description in Figure 2 of the structure and application of an evaluation framework designed to meet the current requirements of CAL researchers can be summarized in four points:

- generation of a detailed description of the evaluation questions to be answered , i.e. the quality of learning outcomes and the means of effective measurement;

- assessment of the evaluation opportunities presented and methods available in the focus situation;

- consideration of findings in relation to the influence of prevailing situational factors;

- reflection on the evaluation process and study findings with a view to future actions.

CAL is not used alone, and so should not be evaluated in isolation. Attempts to do this have frequently been related to measures of cost-effectiveness or comparative studies. Without minimizing the importance of these issues, they do not address the critical questions about quality of learning and the integration of CAL.

To end on an optimistic note, it was once said of the Model T Ford that if proof had been needed that the motor car provided an economical form of mass transport, it would never have passed the novelty stage. What mattered in the end was that it increased user-choice and provided an enjoyable, effective way to travel, so the economics of production became a priority and eventually made it affordable to the masses. Cost considerations of CAL involve a rather separate and complex set of issues, and no attempt is made to include them in the SECAL framework. The primary focus is on how CAL technology might enhance the quality of learning outcomes in the short term, and in the longer term help to drive major educational and social change.

References

Angus, L. (1986), 'Developments in ethnographic research in education: from inter-pretative to critical ethnography', *Journal of Research and Development in Education*, 20 (1), 23–31.

Bates, T. (1981), 'Towards a better research framework for evaluating the effectiveness of educational media', *British Journal of Educational Technology*, 12 (3), 215–33.

Brown, J. S., Collins, A. and Duguid, P. (1989), 'Situated cognition and the culture of learning', *Educational Researcher*, 18 (1), 32–42.

Carr, W. and Kemmis, S. (1986), *Becoming Critical: Education, Knowledge and Action Research*, London and Philadelphia: The Falmer Press.

Darby, J. (1992), 'Computers in teaching and learning in UK higher education', *Computers in Education*, 19 (1–2), 1–8.

Dede, C., Fontana L. and White, C. (1993), 'Multimedia, constructivism and higher-order thinking skills', in Maurer, H. (ed.), *Educational Multi Media and Hypermedia*, Boston MA: AACE.

Denzin, N. K. and Lincoln, Y. S. (1994), *Handbook of Qualitative Research*, London: Sage.

Draper, S. W., Brown, M.I., Henderson, F. P. and McAteer, E. (1996), 'Integrative evaluation: an emerging role for classroom studies of CAL', *Computers and Education* (CAL '95 Special Edition).

Ellis, D. (1996), 'Learning with technology (Australia)', Queensland University of Technology via AAHESGIT Electronic Mailing List (LISTPROC@LIST.CREN.NET), archived posting of 1 March 1996.

Elton, L. R. B. and Laurillard, D. M. (1979), 'Trends in research on student learning', *Studies in Higher Education*, 4 (1), 87–102.

Emery, D. (1993), 'Developing effective instructional graphics', *Journal of Interactive Instruction Development*, Fall, 20–4.

Fontana, L., Dede, C., White, C. and Cates, W. (1993), 'Multimedia: a gateway to higher order thinking skills', in Maurer, H. (ed.), *Educational Multi Media and Hypermedia*, Boston MA: AACE.

Geoghegan, W. (1994), 'Whatever happened to instructional technology', paper presented to the 22nd International Conference of the International Business Schools Computing Association, Baltimore, USA. This paper is available at: *http://w3.scale.uiuc.edu:80/ scale/library/geoghegan/wpi.html.*

Glaser, R. and Strauss, A. (1967), *The Discovery of Grounded Theory*, Chicago: Aldine.

Gunn, C. (1995), 'Usability and beyond: evaluating educational effectiveness of computer-based learning', in Gibbs, G. (ed.), *Improving Student Learning Through Assessment and Evaluation*, Oxford: Oxford Centre for Staff Development.

Gunn, C. (1996), *A Framework for Situated Evaluation of Learning in Computer Environments*, Ph.D. thesis, Institute for Computer Based Learning, Heriot-Watt University.

Gunn, C. (1996 in press), 'CAL evaluation: what questions are being answered?', *Computers and Education*, 27 (4).

Gunn, C. and Maxwell, L. (1996), 'CAL in human anatomy', *Journal of Computer Assisted Learning* (in press).

Hammond, N., Gardner, N., Heath, S., Kibby, M., Mayes, T., McAleese, R., Mullings, C. and Trapp, A. (1992), 'Blocks to the effective use of information technology in higher education', *Computers and Education*, 18 (1–3), 155–62.

Jonassen, D. H. (1993), 'The future of hypermedia-based learning environments: problems, prospects and entailments', in Maurer, H. (ed.), *Educational Multi Media and Hypermedia*, Boston MA: AACE.

Keller, J. (1987), 'Strategies for stimulating the motivation to learn', *Performance and Instruction*, 26 (8), 1–7.

Kemmis, S. (1987), 'Schools computing and educational reform' in Bigum, C., Bonser, S., Evans, P., Groundwater-Smith, Grundy, S., Kemmis, S., McKenzie, D., McKinnon, D., O'Connor, M., Straton, R. and Willis, S., *Coming to Terms with Computers in Schools,* report of the Schools Studies of the National Evaluation Study of the Commonwealth Schools Commission's National Computer Education Program, Deakin Institute for Studies in Education, 289–306.

Lybeck , L., Marton F., Stromdahl, H. and Tullberg, A. (1988), 'The phenomenography of the mole concept in chemistry', in Ramsden, P. (ed), *Improving Learning: New Perspectives*, London: Kogan Page.

MacDonald, B. and Jenkins, D. (1979), *Understanding Computer Assisted Learning*, Norwich: University of East Anglia.

Mayer, R. E. and Anderson, R. B. (1991), 'Animations need narrations: an experimental test of a dual-coding hypothesis', *Journal of Educational Psychology*, 83 (4), 484–90.

Paivio, A. (1986), *Mental Representation: A Dual Coding Approach*, Oxford: OUP.

Ramsden, P. (1988), *Improving Learning New Perspectives*, London: Kogan Page.

Sorge, D. H., Russell, J. D. and Weilbaker, G. L. (1994), 'Implementing multimedia based learning: today and tomorrow', in Reisman, S. (ed.), *Multimedia Computing: Preparing for the Twenty-first Century*, Harrisburg and London: Idea Group Publishing.

Spencer, K. (1991), 'Modes, media and methods: the search for educational effectiveness', *British Journal of Educational Technology*, 22 (1), 12–22.

Steinberg, E. (1989), 'Cognition and learner control: a literature review 1977–88', *Journal of Computer Based Instruction*, 16 (4), 117–21.

Yin, R. (1991), *Case Study Research: Design and Methods*, London: Sage.

Zuber-Skerrit, O. (1990), *Action Research for Change and Development*, Brisbane, Australia: CALT, Griffith University.

Update – CAL evaluation: future directions

Since publication of the SECAL Framework in 1996, the use of technology in education has advanced rapidly on many fronts, bringing challenges as well as positive outcomes. Many of the emergent issues have been identified through the application of SECAL and similar evaluation frameworks in different educational settings (Draper *et al*, 1996; Harvey, 1998). The importance of authentic study environments and evaluating the context of use has been confirmed many times, and situational adjustment rather than modification of learning resources has been a recommended course of action following evaluation studies.

Study findings have made a significant contribution to contextual improvements as well as to discipline-specific educational design:

- implementation of institutional policies and practices that support innovation;
- increasing levels of fluency with technology through training and development;
- improved educational software design, functionality and availability;
- design of Rich Environments for Active Learning with integrated technology applications.

Feedback on positive and negative influences through the relevant organizational channels, e.g. IT Services, policy development committees or faculty boards, allows appropriate action to be taken. Ideally, an action research model involves representatives of all stakeholder groups in an evaluation team. In reality, the process of change may be less efficient.

Evaluation of teaching and learning innovations has also produced evidence of the educational potential of different types of technology as the growing body of literature demonstrates. For the most part, this has not been a result of experimental studies.

Indeed, the No Significant Difference literature (*http://cuda.teleeducation.nb.ca/nosignificantdifference/*) reveals insufficient benefits to justify the investment of time and resources. Because the benefits of technology in education are increasingly obvious, it must be concluded that the methods, in these cases, are flawed. The popular evaluative and situated approaches provide sufficient evidence to support grounded theory development on issues such as:

- the benefits of animation and simulation for demonstrating complex processes;

- the use of drill and practice exercises to reinforce learning and build confidence;

- the positive impact of computer-mediated communication on learning;

- the use of multimedia to support different learning styles and engage user interest.

These approaches have proved instrumental in the development of effective uses of Internet technology, flexible learning and innovative teaching methods. In such a rapidly advancing field, it would be difficult indeed to hypothesize and measure largely predictable outcomes.

References

Draper, S. W., Brown, M. I. *et al* (1996), 'Integrative evaluation: an emerging role for classroom studies of CAL', *Computers and Education*, 26 (1–3), 17–32.

Harvey, J. E. (1998), *The Evaluation Cookbook*, Learning Technology Dissemination Initiative, Edinburgh: Heriot-Watt University.

Contact author

Cathy Gunn is leader of the Education Technology Program in the Centre for Professional Development at the University of Auckland. Her research interests are integration and evaluation of technology in higher education, flexible learning principles and practice, strategic planning and institutional change management and electronic democracy. [*ca.gunn@auckland.ac.nz*]

Institutional change

Enabling learning through technology: some institutional imperatives

Audrey McCartan, Barbara Watson, Janet Lewins and Margaret Hodgson
Education and Training Area, IT Service, University of Durham

Initially published in 1995

This paper considers the importance of the institution as the dynamic interpretative element on which will depend the successful integration of the learning technology developed through our national initiatives into the academic curricula of Higher-Education institutions. Based on our experience of working on teaching technology programmes, within the framework of national and institutional initiatives, it is evident that the establishment of an institutional strategy, and its implementation in a supporting university-wide programme of staff development and training, together with strong commitment at the senior managerial level, are imperatives which determine the successful integration of learning technology within academic institutions.

Introduction

The imminent completion of many Teaching and Learning Technology Programme (TLTP) projects means that a considerable number of courseware deliverables will soon be available to Higher Education (HE) institutions. The Higher Education Funding Council's intention in funding the Programme (HEFCE Circulars, 8/92, 13/93) was to ensure their integration into academic curricula by providing institutions with an opportunity to review their 'teaching and learning culture' with regard to the embedding of learning technology within their institutional practice. Two recent workshops, conducted with a representative sample of newly appointed academic staff in connection with the evaluation of materials to be included in a staff development pack whose purpose is to encourage the use of IT in teaching and learning (TLTP Project 7), strongly suggested that the availability of courseware alone was insufficient to ensure its integration into educational practice. The establishment of enabling mechanisms at the institutional level, as well as within departments, was crucial to ensure the effective use of learning technology.

Institutional commitment

The priorities of the national teaching technology initiatives which have been funded during the last decade (CTI, ITTI, TLTP) have reflected the increasing capacity and functionality of hardware and software as well as the growing emphasis on curriculum development. There has been a development from computer-led initiatives to pedagogically driven priorities. It is within the institutional context that these nationally identified objectives, institutional policy and implementation meet, where policy is developed and programmes delivered. The institution is the dynamic interpretative element working between the national and local levels, yet to date the institutional perspective has been the least regarded.

In a recent ITTI project, which investigated the core IT skills which were required by university staffs in order to execute their job functions, the institutional context, particularly the strength of institutional support, was found to be the most significant factor influencing implementation and practice. The report (Hodgson *et al*, 1994) was the result of a nationwide consultation exercise in which about a quarter of UK universities participated. Initial advisory discussions were conducted with representatives of the variety of interests which contribute to the formulation of IT and teaching and learning policy – IT managers, staff development advisers, teaching and learning co-ordinators, and senior managers. Sixteen universities, drawn equally from the former UFC and PCFC sectors, went on to form internal working parties to identify which core IT skills were required within their institutional context amongst academic, administrative and clerical staffs, thus enabling them effectively and efficiently to operate their job functions consistently with institutional priorities. Each institution collected relevant background information about itself, then defined and conducted the business of its working party in order for the results to be most informative for its own forward planning as well as reporting back to the project.

The institutional reports indicated that there was an already pervasive use of information technology across institutions and that its application in all areas of a university's business was rapidly increasing. Generally, staff had a positive attitude to the use of information technology, and expected to be offered the opportunity to develop their skills further. Current training provision was, however, often patchy and uncoordinated and was likely to be supply- rather than needs-driven. It was not always related to institutional needs.

Effective information technology training provision was most likely to arise from firm commitment at the institutional level. The foundation of this commitment was the development of an institutional information strategy with associated faculty or departmental strategies. These strategies would incorporate responsibilities for the installation of up-to-date technology, made widely available across distributed networked systems, together with the provision of appropriate training opportunities. There needed to be a clear relationship between institutional strategy, staff development and training programmes. It was essential for this relationship to be reflected in effective resource allocation mechanisms, especially in a devolved budgetary organization.

For the delivery of training programmes, a structured approach was preferred. This might be based on the identification of core skills, related to particular job functions. A 'bricks' approach, where the skills were defined in terms of levels, would enable training provision to link the acquisition of particular skills and levels to specified institutional functions. The strong staff development interest would ensure coherent and co-ordinated provision for individual members of staff while supporting the effective implementation of institutional planning.

In order to ensure responsiveness to user needs, the programme would encompass a variety of learning opportunities. Although traditionally these have mainly been provided through taught courses with supporting documentation, new learning methods and mixed media were now available to enable increased flexibility and customization in delivery. The recognition of informal methods of learning, such as training selected staff to undertake limited training responsibilities within their section, would maximize the use of relevant expertise and experience.

An identified group of staff for whom particular provision was urgently required was senior managers. They were responsible for the formulation of policy and its implementation. They needed to be aware of, if not actually competent in the use of, the wide ranging power of modern computer networks and its impact on institutional practices, as well as the implications of the pervasiveness of technology for future planning, including job functions (TLTP Project 45).

Teaching Technology Initiatives

The institutional imperative emphatically to support staff development within the implementation of a university-wide policy to encourage, support and develop the use of information technology in teaching and learning has been implemented at the University of Durham through two Teaching Technology Initiatives posts. These staff work in partnership with academics to review their course structure, content, delivery, assessment and evaluation methods in respect of the appropriate and effective use of IT. They co-ordinate the dissemination of knowledge and increase awareness about developments in computer-aided learning. The awareness program includes university-wide and departmental presentations and hands-on sessions, staff development sessions, contributions to text-based newsletters, and running mail groups. Development work in departments includes advice, information and the development of substantial courseware projects. To date, courseware has been developed successfully for projects in archaeology, chemistry, education, geography and Spanish.

Selecting software for successful courseware

The success of courseware in achieving educational objectives is in part dependent on selecting the most appropriate development tool. The learning objectives must be analysed, then clearly defined in order to evaluate which software would be most suitable to achieve these educational goals.

Consequently, at the University of Durham, courseware has been developed in:

- the core spreadsheet application – *Quattro Pro for Windows*

- the assessment package – *Question Mark Professional*

- the authoring package – *ToolBook*

Learning Errors in Measurement in Chemistry was developed using the University-recommended spreadsheet, *Quattro Pro for Windows*, to utilize its spreadsheet functions. The first-year chemistry students start by acquiring a foundation knowledge of Quattro Pro, using the self-study guides of the Computer Literacy Programme. The students need to understand how to navigate through Quattro Pro in order to be able to use the full potential of the courseware. They then use the CAL material in supervised practical sessions as an integrated part of their course.

Students read the explanatory text, and are then directed to pages of exercises. They learn interactively by calculating their own values for the exercises and entering them into the spreadsheet. The built-in spreadsheet functions are used, through macros, to make the graphs change according to their values and to check their answers. The students click the Am I Right? button to find out if their answers are correct or need modifying. The courseware is based on random numbers, so the students all have different values for their exercises, and are able to start again with new values if they wish. There is a summary page which holds all the values entered by students and the students are requested to give a hard copy of this page to the lecturer.

Question Mark Professional is an easy-to-use computerized assessment and examination program. Its strength lies in its analysis and reporting facilities. As part of the preparation for a departmental test, a General Knowledge Test has been written to enable the

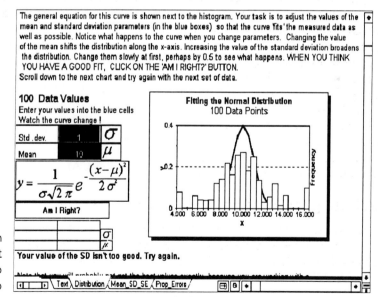

The general equation for this curve is shown next to the histogram. Your task is to adjust the values of the mean and standard deviation parameters (in the blue boxes) so that the curve 'fits' the measured data as well as possible. Notice what happens to the curve when you change parameters. Changing the value of the mean shifts the distribution along the x-axis. Increasing the value of the standard deviation broadens the distribution. Change them slowly at first, perhaps by 0.5 to see what happens. WHEN YOU THINK YOU HAVE A GOOD FIT, CLICK ON THE 'AM I RIGHT?' BUTTON.
Scroll down to the next chart and try again with the next set of data.

Figure 1: Errors in Measurement developed in Quattro Pro

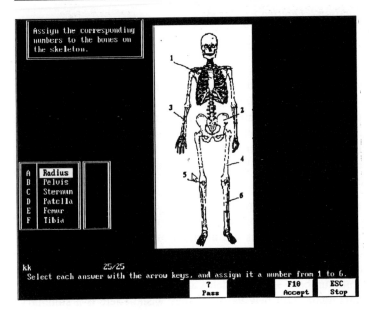

Figure 2: General Knowledge Test – Question Mark Professional

students to attempt the different question types available in *Question Mark*. At the end of a departmental test, the students were asked if they found the software easy to use, and 97 per cent found it 'Easy' or 'Very easy'. This positive result is partly due to the students being able to become familiar with the package through the General Knowledge Test.

Until recently, Spanish syntax had been taught using only lecture notes. Students had found this topic difficult to grasp, and the aim of the *Spanish Syntax* project is to use the hypermedia facilities of *ToolBook* to enhance the learning of the subject material through interactive examples and tests. Colour is used consistently throughout the courseware to identify the different syntactical units of sentence structure.

There are different types of exercises, including tree diagrams, following each explanation of the syntax. In many of the exercises, the students click the Am I Right? button to find out if they have completed the exercise correctly. If they have not answered correctly, they are given hints, e.g. on vocabulary or abbreviations. After two attempts, the answer is given to them. This is to allow the students to be able to learn through their mistakes and progress to the next question. The students use this CAL material as an integrated part of their course.

Conclusion

Firm institutional commitment is imperative to enable learning through technology. A clear relationship between an information technology strategy, staff development and training provision, supported by appropriate resource allocation mechanisms, is essential to ensure coherent and co-ordinated provision for all staff.

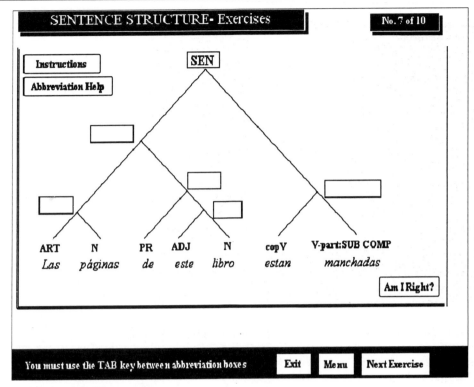

Figure 3: Spanish Syntax – ToolBook

Selecting the most appropriate software for the development of courseware has resulted in successful and enjoyable learning. All the courseware and computer-based assessment is delivered on the campus-wide Novell network, so the students are able to progress at their own pace and repeat sections and exercises.

References

Computers in Teaching Initiative (1984), University Grants Committee.

Core IT Skills for Teaching and Learning in Higher Education: Tools for the Development of a National Framework, TLTP Project No. 45.

Hodgson M., McCartan A. and Hare, C. (1994), *A Framework for the Development of Core IT Skills for University Staffs*, CVCP/USDU Publications.

Information Technology Training Initiative (1991), Universities' Funding Council.

IT In Teaching and Learning: A Staff Development Pack, TLTP Project No. 7.

Teaching and Learning Technology Programme Phase 1 (1992), HEFC(E), Circular 8/92, March.

Teaching and Learning Technology Programme Phase 2 (1993), HEFC(E), Circular 13/93, April.

Update – Enabling learning through technology: some institutional imperatives

The last five years have seen an increase in institutional commitment to learning technology through the creation of a set of 'new faces' to support its use. Learning development officers, working under various titles, have been appointed with institution-wide briefs to integrate the efficient and effective use of C&IT into the curricula of higher education institutions.

The remarkable development of the Internet has allowed students and staff to access a wealth of online resources and has opened up distance learning across the globe. Increasingly, quality courseware is becoming available. Students have generally recognized the benefits of C&IT; in contrast some staff have been slower to see these benefits, perhaps in part through a lack of C&IT skills, but also through a fear of the wider implications.

The combination of these new specialists to support the use of learning technologies and the increase in the range of resources available might suggest a natural improvement in the student learning experience. And indeed there are some examples of good practice where learning technologies have enriched the quality of the learning experience by, for example, simulating real-life situations or by encouraging student collaboration and online debate. However, unfortunately this is not universally true.

The fundamental issue is the degree of readiness of academic staff to capitalize on these opportunities. Integration of C&IT into the curriculum requires a change in curriculum development, and needs dedicated time to be effective. This integration rests on a willingness to change the course, rather than simply bolting the technology on to the course. The issue of the Not Invented Here syndrome is still present. Therefore, effective integration is likely to need nothing less than a new course, starting with new aims and objectives, different learning and teaching methods and new means of assessment and evaluation. All this takes time. An associated difficulty is the real (or perceived) un-reliability of institutional networks to deliver courseware.

Although students can now gain much by using material freely available on the Web or through the university networks, this falls short of the dynamic, creative, enhancing learning experience in which there is a seamless interface of integrated learning resources, with technical and non-technical materials complementing each other.

Principal author

Audrey McCartan was formerly Arts Computing Development Officer at the University of Exeter and a Lecturer in Information Technology at the University of Sheffield. She was Manager of the Computer Literacy Programme and Director of the Teaching Technology Initiatives Group at the University of Durham.

Reinventing the university

Stephen Brown
De Montfort University

Initially published in 1998

In view of the discernible trends in Higher Education towards increased numbers, increased diversity and expectations of students, declining resources, increased competition between providers and increasingly irregular patterns of student attendance, there is a need to establish new ways of delivering and supporting teaching and learning. The convergence of different media into a single networked digital domain, and the rapidly expanding accessibility of such media, both driven by market forces, is tempting many HE institutions to develop an online presence. True 'virtual universities', i.e. completely online, are still the exception and there are strong reasons why traditional, face-to-face, universities should resist the pressure to develop a virtual alter ego. Nevertheless, online delivery and support of teaching and learning offers many potential benefits to students of such institutions, not least of which could be freedom from the constraints of conventional attendance patterns.

Since significant new resources are unlikely to be available to most UK institutions to develop online operations, existing internal resources need to be re-engineered to bring them into line behind the chosen venture. Small scale, bottom-up experiments are unlikely to succeed in re-engineering sufficient resources to achieve lasting institution-wide change. Re-engineering the campus requires a culture change, which needs to be led from the top, through an unbroken chain of champions down through the organization, including the resource managers at middle management level. A minimum of two years needs to be allowed for evaluation of the impact of change. During that initial period central co-ordination of, and support for, individual projects is essential to ensure delivery against targets of time, cost and quality. Nevertheless institutional change can best be achieved by encouraging local ownership, which means allowing local autonomy and avoiding premature imposition of standards.

Change

Higher Education in the UK and elsewhere is under considerable pressure to change as a result of:

- rapid growth in student numbers
- increased variability of student intakes
- declining resources
- concerns about quality of graduates
- increased pressures for public accountability
- concerns about the relevance to industry
- increased competition from other countries and commercial interests
- increased availability and reduced costs of networked media (Brown, 1997; Cunningham *et al.*, 1998).

These factors are driving universities to reinvent themselves. Traditional behavioural models of education are giving way to more learner-centred, collaborative experiences (Hardy, 1997). Learner-centred approaches accommodate learner needs and acknowledge that learners may have other calls on their time, leading to more flexible delivery and support strategies which allow learners to shift the time, place and pace of learning. Network technologies such as the Internet, email and computer-based and video-conferencing can assist with this process and at the same time they open up opportunities for learners to seek out and utilize learning resources offered by other parties.

Coping strategies

Complexity in the environment can be reduced by increasing the span of control of the organization (Emery and Trist, 1969). Collaborative ventures such as the University of the Highlands and Islands in Scotland (Cornwell, 1997), or Western Governors University in the USA (Romer, 1997) are examples of a trend towards mega-universities identified by Daniel (1997). 'Virtual' universities such as Western Governors (http://www.wgu.edu), California Virtual University (http://www.california.edu) and Colorado University 'CU Online' (http://www.cuonline.edu) are another way of expanding the span of influence, without requiring a physical presence. However, much of so-called online education still depends on video-conferencing, books, residential schools, etc. (Ebeling and Bistayi, 1997; Philips and Yager, 1998).

Ten good reasons not to go virtual

1. The costs of developing a virtual online presence are not yet fully understood but are considerable. In a climate of diminishing resources, investment capital has to be diverted from existing functions.

2. Competing organizations need to be able to respond rapidly and flexibly to new market requirements and competitive threats. Yet quality learning resource materials and learning support systems take longer to develop than traditional teaching.

3. Classroom teaching can be modified much more readily than resource-based learning with its longer development timescales.

4. The infrastructure requirements of the virtual institution are quite different from those of traditional institutions and require considerable capital investment.

5. Although there is evidence that the economics of purely online education can be made to work, for a traditional university to make the transition to predominantly mixed mode would require that institution to run both systems concurrently, with all the associated costs and inevitable tensions.

6. Staff attitudes to curriculum change on this scale are often a significant barrier (Brown, 1997).

7. Student needs: not all learners are comfortable with independent learning, particularly among the traditional 18–22-year-old cohort of established universities (Hutton, 1998).

8. Reallocation of resources could weaken the traditional side of the operation at a time when competition is increasing.

9. It can be argued that there are no blueprints for converting an existing university to online mode on an institutional scale, or at least, knowledge of how to do so has not yet been unequivocally demonstrated.

10. Traditional universities have strengths that are lost in cyberspace: physical locations that provide a respite from the everyday demands of home and work; facilities and equipment that are not readily simulated in virtual space, people with whom to socialize, enjoy physical contact, chance encounters, etc.

An Electronic Campus

Instead of developing a virtual presence, De Montfort University has decided to create an Electronic Campus, in order to:

1. Enhance flexibility of access to learning on campus.

2. Develop a position that can lead to collaborative ventures with other organizations.

During 1997/8 the goal has been to:

- Develop a substantial body of HTML compliant teaching, learning and assessment materials, which can be accessed readily by students on campus and traded with other institutions.

- Put in place appropriate learning support systems that encourage and facilitate peer group interaction and staff-student exchanges, independent of time and specific university location.

- Achieve a rapid return on investment of resources.

- Re-engineer existing resources to achieve the above.

The Electronic Campus is employing a combination of three possible implementation strategies:

- In-house development.

- Embedding of materials developed elsewhere.

- Collaboration with other organizations to develop jointly new materials.

Pump-priming funds have been made available centrally for faculties to bid against. A proforma guides faculty staff through a set of questions designed to encourage them to consider key issues and address the selection criteria applied to proposals.

Staff IT training courses have been restructured to provide basic technical authoring skills using Microsoft Office products and more advanced Internet training on HTML authoring, web site development, Internet communications and information search techniques. Project teams are also offered short courses on the use of specific software applications used within the Electronic Campus.

Impact

Faculty	Projects
Applied Sciences	6
Art & Design	5
Business & Law	3
Computing & Engineering	3
Learning Development	3
Health & Community Studies	3
Humanities & Social Sciences	4
Total	27

Table 1: Distribution of faculty projects

So far there are 27 different Electronic Campus projects, ranging across all faculties and at all levels including FE, undergraduate, postgraduate and continuing professional development. Some of the projects are complete courses, some are full modules and others are parts of modules. The first of these projects will start delivering to students during the first semester of 1998.

Timescale

From the initial decision to establish an Electronic Campus (June 1997) through to delivery of the first products (October 1998) has taken 16 months. Key milestones on the way are outlined below.

July 1997: Establishment of a project director. Individual briefings between the project director, the Pro Vice-Chancellor with overall responsibility for the initiative and each of the heads of the 14 academic schools, to explain the goals and invite project proposals.

September 1997: First proposals received. Establishment of a formal steering committee to formulate policy and strategy.

During 1997/8 the 14 schools were reorganized into 6 faculties and in January 1998 a central team was put in place to interface with the new faculties. The role of these Learning Development Managers (LDMs) has been to help specific faculties to:

- Develop appropriate teaching, learning and assessment strategies.

- Develop teaching and learning proposals for internal and external funding.

- Identify and obtain resource-based learning materials produced elsewhere.

- Develop and implement learning support strategies and systems.

- Identify and meet staff development needs in relation to implementation of the Electronic Campus.

Successful project bids result in the formation of Project Teams within faculties. LDMs provide ongoing assistance and a link back to the central production resources (graphic design, courseware authoring, programming, desktop publishing, audio, video and photography, digitizing, print, etc.).

By April 1998 it was apparent that there was a need for an intermediate level of management between the Steering Group and the LDMs. An Operational Group of senior managers with responsibilities for key functions such as libraries and IT infrastructure was created to determine implementation uses and strategies.

In May 1998 a Web site was launched to support and publicize the initiative in conjunction with internal newsletter articles (13 to date).

September 1998: Hardware and software standards agreed.

Costs

It is not easy to cost a venture such as the Electronic Campus because it is not being developed *ab initio*. A lot of the costs would have been incurred anyway by the institution as part of its normal running costs. The following figures are an attempt to tease out some additional costs directly attributable to the initiative.

Not surprisingly the infrastructure costs are a significant element. A figure of £400 per student includes an allowance for network hardware and software installation, servers, access terminals and technical support. It is assumed for the purposes of the calculation that the ratio of terminals to students is 1 to 5 and that all staff have access to their own desktop machine.

The other significant costs are for learning materials and support systems development, including the costs of bought-in software. On average, each project costs around £33,000, split on a 60:40 ratio between faculties and the centre.

A cost that is easy to overlook is the amount of time spent by senior staff determining policy strategy and implementation plans. It is estimated that the Electronic Campus has cost approximately £8,000 p.a. so far pro rata.

Problems

As the number of projects under way has increased, it has been increasingly difficult to keep track of progress and to maintain the level of input necessary to guarantee minimum standards of quality. This highlights a trade-off to be made between embedding in local practice and quality control.

Inevitably there have been barriers to progress, which in some cases have resulted in slippage against project schedules. This experience highlights the need to retain some flexibility in delivery methods.

There are conflicting pressures for and against the early adoption of standards for hardware, development tools, delivery software, conferencing software, etc. Standards can reduce costs, staff development needs and product development times. Offset against this is the risk of committing too early to inappropriate tools and resistance to standards by staff interested in pursuing different avenues of development. This highlights the need to balance central co-ordination against the need for local autonomy and to postpone making commitments to standards until there is a high degree of confidence in their effectiveness and relevance.

There is also the potential for conflict with other organizational standards, for example, computing policies designed to maximize network integrity and data security may be at odds with the need for open and flexible access to learning materials, assessments and personal performance records. This highlights the need to ensure close collaboration between Electronic Campus developments and relevant operational arms of the university at the highest possible management levels. 'Bottom up' implementations are unlikely to be successful on an institutional scale.

The resource re-engineering model adopted here depends on being able to realign an increasing proportion of existing resources behind the new activities. Without convincing evidence of the benefits to staff, students, departments and the institution as a whole, it is very difficult to obtain the support required at resource manager level to sustain development. Evidence of this kind takes time to collect and present. This highlights the need for senior management commitment over an initial development and implementation phase of at least two years.

Lessons learned

Money is not enough
Aside from the infrastructure costs, the major Electronic Campus investment has been staff time, as buyout of staff from other normal duties and recruitment of additional staff. Money is not enough however to ensure adequate human resources. In our experience the faculty staff taking a lead in Electronic Campus have typically been heavily committed to other key activities as well, such as student recruitment, research, university consultancy, administration, etc. It has been difficult for such people to allocate as much time to Electronic Campus projects as they themselves would like. Other rewards need to be put in place such as recognition of the importance of teaching innovation and excellence.

Champions at all levels
Bottom-up approaches tend to founder on the rocks of competing policies and standards central to different parts of the institution. Staff eventually move on to other things and business returns to usual practices. On the other hand, a top-down, management-led, approach can be frustrated at the level of middle management where hard choices have to

be made about resource allocations in the face of strong competing pressures. In order for innovation successfully to permeate the entire organization there have to be champions at all levels, ensuring an unbroken chain of commitment to the vision (Brown, 1997).

Faith versus evidence

Experiments need faith in the probability of desirable outcomes. On the other hand, wholesale institutional adoption of change needs convincing evidence. It takes time to build the evidence in a given organizational context and so experiments have to be given a reasonable period in which to prove themselves. In our experience, two years is an absolute minimum period for development and trialling.

Central co-ordination versus local autonomy

Top-down management is essential to ensure central co-ordination at a high enough level within the organization to ensure that inter-departmental differences do not present impossible barriers to innovation. Yet without a strong sense of local ownership in the faculties themselves, innovation is unlikely to be pursued wholeheartedly or sustained in the long term. Commitment is most likely to stem from local control over decision-making, so the temptation to control everything from the centre needs to be resisted.

Standards versus flexibility

From this it follows that while standards can be helpful in the early stages, they should not be imposed at the expense of constraining flexibility and creativity on the part of those expected to develop and implement the innovation. As the innovation gathers momentum and supporters, the need for greater co-ordination and agreement on standards can be allowed to emerge collectively.

Keep it simple

A university is a complex organizational form, which has evolved over hundreds of years. The Electronic Campus has had to be made operational within a year. On this timescale it has been important to make everything as easy as possible for the staff and students involved. The simplest, most familiar and most readily available tools have been selected to facilitate this process.

References

Brown, S. (ed.) (1997), *Open and Distance Learning: Case Studies from Industry and Education*, London: Kogan Page.

Cornwell, T. (1997), 'The University of the Highlands and Islands', paper for Using the Virtual University to Develop Capability, Higher Education for Capability Conference, Leeds, 27 June 1997.

Cunningham, S., Tapsall, S., Ryan, Y., Stedman, L., Bagdon, K. and Flew, T. (1998), 'New media and borderless education. A review of the convergence between global media networks and higher education provision', Department of Employment, Education, Training and Youth Affairs: Australia. *http://www.deetya.gov.au/divisions/hed/highered/eippubs/ eip97 22.*

Daniel, J. (1997), *Mega-Universities and the Knowledge Media: Technology Strategies for Higher Education*, London: Kogan Page.

Ebeling, A. and Bistayi, S. (1997), 'Wired degrees: Forbes' 20 top Cyber-Us', *http://www.forbes.com/forbes/97/0616/5912084a.htm.*

Emery, F. E. and Trist, E. L. (1969), 'The causal texture of organizational environments', in Emery, F. E. (ed.), *Systems Thinking*, London: Penguin.

Hardy, D.W. (1997), 'Instructional design for distance education', *Open Praxis*, 1, 26–9.

Hutton, J. L. (1998), 'What's the difference?', *Open Praxis* 1, 19–21.

Philips, V. and Yager, C. (1998), 'Best distance learning graduate schools: earning your degree without leaving home', *Princeton Review* Random House. See also *http://www.geteducated.com/bestgrad.htm.*

Romer, R. (1997), 'A matter of degrees', *Educom Review*, January/February, 17–23.

Update – Reinventing the university

The Electronic Campus project has successfully delivered 200 modules to 3,000 students across all faculties of the university. Staff who were initially cautious have become enthusiastically committed to the point where the amount of time they spend greatly exceeds that devoted to development and delivery of face-to-face teaching. Notwithstanding some minority views and experiences, on the whole students are positively predisposed towards online learning, their expectations have been largely met and in some cases actually surpassed. More objectively there has been no significant difference in student assessment results for those working on Electronic Campus modules, compared either with peers studying traditionally or with previous cohorts on the same modules. However, the down side of the policy decision to keep it separate from mainstream university processes in order to minimize institutional risk has been the marginalization of Electronic Campus. A successful strategy should address the process of obtaining commitment as well as generate course content. The tentative conclusion here supports the notion that a ring-fenced project is a good strategy to establish rapid results and momentum (Carey *et al*, 1999), but that in the longer term it can have only limited impact.

The next challenge is to scale up the initiative to establish it as mainstream activity. This is being tackled through curriculum planning and resource allocation, quality assurance procedures, staff development policies and priorities and staff reward systems. All faculties have to generate strategic development plans for learning and teaching, identifying their primary student target populations, learning and teaching methods, resourcing plans, staff and student development requirements and infrastructure needs. University quality assurance procedures have been revised to ensure that new or revised modules and courses take account of changes in pedagogical philosophies and technologies. A new Centre for Learning and Teaching has been created to lead on staff development and accreditation and to run a Teacher Fellowship scheme that rewards innovation and excellence in teaching. Teacher Fellows have the same academic status as Readers, but their role is to encourage good practice in learning and teaching and act as champions for change in their faculties. Finally, online learning resources are being linked to other university information systems to create an integrated learning support environment that is individual to each learner.

References

Carey, T., Harrigan, K., Palmer, A., Swallow, J. (1999), 'Scaling up a learning technology strategy: supporting student/faculty teams in learner-centred design', *ALT-J*, 7 (2), 15–26.

Contact author

Professor Stephen Brown is currently Head of Learning Technologies, Director of the International Institute of Electronic Library Research and Manager of Continuing Vocational Education at De Montfort University. He has led the development of the Electronic Campus at De Montfort University. [*sbrown@dmu.ac.uk*]

Supporting organizational change: fostering a more flexible approach to course delivery

Gail Hart, Yoni Ryan and Kerry Bagdon
Teaching and Learning Support Services, Queensland University of Technology

Initially published in 1999

Queensland University of Technology (QUT) adopted a flexible delivery policy in 1996. The main objective of the policy was to develop a more student-centred approach to teaching and learning, since QUT's student population is predominantly part-time, 'mature age', already in employment, and very diverse in cultural and academic background. For many staff, the policy was threatening: staff were uncertain where they might begin to adapt their traditional face-to-face teaching approaches to overcome the limitations associated with time and place, and they were fearful that their teaching role and academic expertise might be superseded by a technological alternative. They lacked confidence to incorporate appropriate and relevant technologies in an innovative and effective way to support student learning objectives. This paper focuses on the implementation of QUT's policy. It highlights the role of a central services department, Teaching and Learning Support Services (TALSS), in providing training and fostering cultural change across the university. The implementation was guided by a model of flexible education and a set of principles underpinning a 'whole of organization' approach to flexible delivery. Strategies for supporting innovators, sharing experience across disciplines, co-ordinating and focusing the support of educational developers, and embedding staff development processes are outlined. The success and limitations of the organizational change strategy are summarized as 'lessons learned' to inform ongoing institutional policy and procedures.

Organizational contexts

In Australia, as in the UK, the political, economic and technological contexts within which higher education operates are shifting. Many commentators (e.g. Tiffin and Rajasingham, 1995; Oblinger and Rush, 1997) have predicted that the outcomes for teaching and learning environments will be as shattering to the 'bricks and mortar' universities as the long-predicted tectonic plate shift which will swallow San Francisco.

Our vision is less apocalyptic. Indeed, we are convinced from various research projects in which we have participated (Hart *et al*, 1998; Cunningham *et al*, 1998) that the move to student-centred education, if it is genuinely embraced by universities, will result in a slow evolution of existing university environments.

However, there is no doubt that the last decade has already produced many changes within universities worldwide. Classes are larger, student populations more diverse, both culturally and in terms of their preparation for independent learning. With the introduction of tuition fees and supplementary charges for support services such as computer access, students are becoming more discerning 'consumers'. They expect high-quality learning opportunities and better ancillary services. At the same time, public funding has been reduced as governments of mature economies seek to reduce their outlays by privatizing as many of their traditional services as possible. Tertiary education has become a victim of its own success: in moving from an élite to a mass system, it has ceased to become a public investment by the community in its collective future, and has become a 'private good' (cf. Saul, 1997) requiring economic decisions as to its worth to individual students.

Whether we view these changes with delight or dismay, the reality is that we must respond energetically and responsibly to ensure that we continue to foster optimum learning outcomes for our students. Much of the burden of this work falls upon individual academics. But the organizational culture within which teaching and learning occurs plays an increasingly important role in teaching, for several reasons:

- the increasing use of technology in the delivery of learning materials means that institutions have new challenges, e.g. in the provision and/or co-ordination of computer facilities, to ensure access and equity prevail;

- individual academics cannot master alone the intricacies of technology-aided communication;

- communication technologies are changing the nature of personal interactions, with less need for face-to-face communication to resolve course enquiries or provide individualized advice.

Analysis of QUT's student population in the late 1990s revealed a changed demographic to predominantly part-time and mature-age students. Students are attracted to QUT because of its strong industry and professional links, high rates of graduate employment, and reputation for applied research. Such a population seemed to demand more flexible administrative and teaching approaches that could accommodate their part-time status, and their need for highly relevant professional qualifications. QUT is also a multi-campus institution with three campuses in the northern suburbs of Brisbane, creating consequent travel difficulties for staff and students who operate across campuses. Hence the university determined to investigate and if necessary 're-engineer' its structure to accommodate changing student needs and better prepare students for their professional futures. It was recognized that cultural change was integral to any systemic change, as both Bates (1995) and Mason (1998) argue.

The term 'flexible delivery' is now widely used in Australia to describe an educational approach that takes advantage of technology and new procedures (such as automated telephone or online enrolment) to allow students more choice in relation to the time and place of their study experiences, and to bring greater efficiencies to an organization's administrative practices. QUT introduced a flexible delivery policy in 1996.

However, the launch of the policy prompted real concern amongst our staff. It coincided with the first round of job losses in Australian universities and was linked with potential staff retrenchments. The emphasis on unfamiliar technologies alienated staff who were committed to face-to-face teaching, and who feared their hard-won discipline expertise would be de-valued in what they perceived to be an inevitable 'online university'. Our staff are not alone: University of Washington staff, USA, have protested strongly against the 'de-humanization' of teaching via online courses (The Industry Standard, 12 June 1998, at *http://www.thestandard.net/articles/news_display/0,1270,670,00.html*).

However, changing practice was perceived as an imperative at QUT, in order to maintain a competitive edge in the higher education sector and in response to the demands of an increasingly discerning student body. The task facing the institution's management was to develop:

- an organizational climate which diminished fear and encouraged appropriate change;

- a support structure which built on staff strengths, and involved staff in devising their own solutions to the problems of the external environment and the needs of their students;

- a robust system, which could accommodate upgrades, software changes, and increasing bandwidth demands, as well as legal changes;

- a support unit that provided the physical infrastructure and the initial/ongoing technical advice, and specialist academic development staff who could bridge the divide between the 'techos' and the teachers.

A model of flexibility in tertiary education

The first stage of the process was to develop a succinct definition of flexible delivery that might serve to allay the fears of staff by focusing on the goal of meeting student needs, thus aligning staff and organizational goals. The focus on flexible delivery was intended to signal to the university community that the strategy was underpinned by a commitment to incremental rather than radical change. Hence the following definition was developed:

> Flexible delivery refers to the use of a range of strategies and technologies to meet the diverse needs of students regarding the location and time of study. Flexible delivery is applicable to both internal and external students (QUT Office of the Deputy Vice Chancellor, 3 July 1998).

The Deputy Vice Chancellor's policy adviser, in consultation with key senior staff in the university, prepared a discussion paper that was widely circulated and then tabled as a policy document at Teaching and Learning Committee and then University Academic Board. A model of flexible delivery was subsequently developed as a framework within which policy and procedures could be determined. We reiterate that integral to the emphasis on flexible delivery was recognition that flexibility was merely the visible tip of a significant cultural change to the traditional university model. Figure 1 represents the framework.

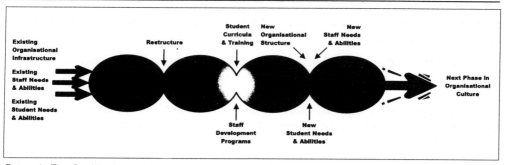

Figure 1: The flexible delivery framework

The structure signals that we see this model as requiring continuous adjustment, not to the extent of the changes of the last two years, but that the intersection of the ovals will always mean deliberate pressure on the organizational structure through management decision-making to support teaching and learning. We recognized that neither flexible learning nor flexible teaching can be mandated. Flexible learning depends on student needs and abilities. Flexible teaching rests on staff abilities such as discipline imperatives, commitment to quality teaching, knowledge of and confidence in using a range of technologies. However, we could, by 're-engineering' our systems and structures, nurture an organizational culture which valued 'learning about teaching and learning in a new context' as its primary goal.

Principles of an organizational approach to flexible delivery

Informing the model and the decisions consequent on the policy were several core principles:

- 'Flexibility' must be linked to specific learning objectives for each course and each subject, based on objective knowledge of student needs and the subject's rationale. While academic judgements regarding these objectives must be the final arbiter of the types of flexibility introduced, academics must also be fully informed of their students' emerging needs, and the potential of technology to assist them in enhancing learning.

- Student skills cannot be presumed in any pedagogic approach drawing on new technologies. Therefore, unlike the MIT Management for the 1990s Framework (Scott Morton, 1991), which focuses on strategy, structure, technology, management processes, and staff skills and roles, the QUT approach includes student skills and roles as a primary component.

- Collaborative approaches drawing on the expertise of academics, technical and learning support staff result in the best possible learning materials and teaching strategies.

- Staff development is embedded in the experience of actually developing flexible options.

- While individual academics must be encouraged to experiment and develop unique materials and strategies, sensible use of central resources means that investment must be largely directed in university-wide or faculty-wide initiatives which contribute to the organization as a whole.

- The 'learning organization' must demonstrate its commitment to whole-of-institution change by recognizing that university culture in particular is strongly individualistic and strongly decentralized. Hence support structures must balance the strengths of centralized units of expertise with the strengths of localized and semi-autonomous units of discipline and teaching knowledge.

- Full evaluation must be integrated in the design and development of all flexible delivery strategies and materials.

Implementation

Once the initial policy on flexible delivery was disseminated throughout the university, a number of activities and processes were undertaken progressively (not always, we should add, in a thoroughly planned way).

- A series of information sessions explicated the changing contexts of higher education, examples and demonstrations of flexible delivery within QUT and in other institutions, and how the administration was proposing to assist staff. These sessions were sponsored by the Senior Executive, thus ensuring that they had a high profile within the university community.

- The policy adviser to the Deputy Vice Chancellor undertook a data collection exercise to determine the extent of existing operational flexibility within the institution, and to focus on the variety of ways 'flexibility' could be perceived and implemented. Currently a more detailed pilot 'mapping' exercise is being undertaken within one faculty to understand better the impact of current initiatives and policy on the development of flexible practices, since we are convinced that information and staff consultation and input are vital to policy development. Changed practice can only result from a sound knowledge base.

- A series of commissioned divisional and departmental reviews over a two-year period from 1996 assessed the effectiveness of the central services supporting teaching and learning, identified potential cost savings and recommended structural changes to improve service provision. A three-staged restructuring process was implemented to forge synergies across previously separate units and ensure a more focused service. The resultant department is a large central unit, Teaching and Learning Support Services, providing a broad range of technical, professional and pedagogical expertise to staff and students.

- A drop-in centre, staffed part-time, was established and featured learning resources that capitalized on technological advances such as data visualization and the range of commercial CD-ROMs available, in a threat-free informal environment outside the formal training programme.

- Since 1992, QUT has provided funding for innovative teaching and learning projects. An internal evaluation of the grant scheme, following an earlier external evaluation, was designed to ensure a closer alignment between the projects funded and strategic university initiatives. The most important result has been our 'large grants', which provide funding up to $150,000 over two years, for projects which embed flexible delivery practices across a faculty, or between faculties. Projects must have the support of the relevant dean or deans.

- Two extensive reports were commissioned from QUT staff with a knowledge of the existing orgainizational culture; one examined the pedagogical dimensions of the new technologies (Rossiter, 1997), the other evaluated existing computer-assisted learning at QUT (Wilss, 1997). The intention was to inform decision-making relating to the contribution that technology could make to increase flexibility.

- The Division invested in infrastructure and initiatives which demonstrated to staff the central commitment to extending flexibility. Among these was a combined student help desk, the student computing guide and the continued development of Datawarehouse as a teaching and learning vehicle as well as a management information system.

- A Web site, 'Teaching and Learning on the Internet', dedicated to teaching with technology in various forms, was established with easy-to-use features, an inviting interface, and links to QUT and international exemplars.

- Specialist positions at senior levels were identified and have progressively been filled with staff who combine both academic and technical expertise; the present authors are respectively at professorial and senior policy adviser (educational media and flexible delivery) levels.

Lessons learned

The key lessons we have learned from the processes and initiatives described above might prove useful to other institutions seeking to avoid wasteful pilot projects and 'wheel reinvention'.

- The visible and energetic support of senior management is critical to any strategic initiative involving cultural change and new approaches to teaching and learning. One of our successful faculties in 1998 admitted that had it not been for the criterion that their proposal for a grant must address flexibility, they would not have had to re-examine the nature of their curriculum and their students' expressed needs, and hence would not have realized how dated their teaching had become.

- Notwithstanding early resistance to the notion of uniting the academic and educational technology support units, it became obvious that unless management 'forced' the issue through amalgamation, the organizational culture would continue to be dichotomized between the 'techos' and the teachers. The resultant unit is a team of general and academic staff who bring their respective skills to catalyse flexible delivery initiatives across and within disciplines and faculty units. The renamed and restructured division (the Division of Information and Academic Services) has activated a high-profile promotion of its staff development activities to complement its existing reputation as a provider of technical and library services.

- Publicizing the successes (and failures) of initiatives through the university avoids *ad hoc* and sporadic small projects which are not sustainable. Workshops, seminars, newsletters, Web discussion groups, training programmes are all vital in this process. The variety of information modes acknowledges the diverse learning approaches of staff and students, and models flexibility.

- Notwithstanding the energy and commitment of the part-time staff member appointed, the drop-in centre was not as successful as we had expected. We surmise that this reflected its part-time status, its confinement to one campus, and the initial attitude of staff.

- The early data collection exercises revealed that 'flexibility' was not clearly and widely understood among staff: for example, many academics reported that they used presentation software such as PowerPoint, with no apparent recognition that computer-developed overheads, of themselves, do not imply flexibility. In order to be used flexibly, overheads must be available on the Web or placed in the library for student access independent of the lecture context in which they are delivered. Such information has informed our staff development activities.

- Restructuring organizations, particularly universities, with their fierce defence of territories, is not without pain for many individuals. Tempering fear and reworking job responsibilities, promoting team responsibility and collaboration while maintaining morale, demand time-consuming negotiation processes which require sensitivity on the part of senior management. Accurate and timely information and communication, we have found, assist to dispel the atmosphere of distrust which can easily arise.

Conclusion

There is no single template for organizational and cultural change within a university, anymore than there is a single template for online delivery. QUT has taken the view that far from inducing cataclysmic change in the shape of universities, the new technologies offer an opportunity for evolutionary development in approaches to teaching and learning, which can accommodate student and staff needs and abilities. That view is informed by a belief that learning in organizations is a continuous and evolving process, and that focusing on the purpose of the university, learning itself, we can develop appropriate institutional responses.

References

Bates, A. (1995), *Technology, Open Learning and Distance Education*, London: Routledge.

Cunningham, S., Tapsall, S., Ryan, Y., Stedman, L., Bagdon, K. and Flew, T. (1998), *New Media and Borderless Education*, Canberra: Australian Government Publishing Service.

Hart, G., Ryan, Y., Williams, H. and Lunney, P. (1998), *Improving the Practice of Mental Health Nursing in Rural and Remote Communities*, A Report for the Rural Health Support, Education and Training Program of the Commonwealth Department of Human Services and Health, Canberra.

Mason, R. (1998), *Globalising Education: Trends and Applications*, London and New York: Routledge.

Oblinger, D. and Rush, S. (eds.) (1997), *The Learning Revolution: The Challenges of Information Technology in the Academy*, Bolton, MA: Anker.

Rossiter, D. (1997), *The Digital Edge: Teaching and Learning in the Knowledge Age*, Brisbane: QUT.

Saul, J. R. (1997), *The Unconscious Civilization*, Ringwood: Penguin.

Scott Morton, S. (ed.) (1991), *The Corporation of the 1990s: Information Technology and Organizational Transformation*, New York: Oxford University Press.

'Teaching and Learning on the Internet', *http://www.tals.dis.qut.edu.au/tlow/tlow.htm*.

Tiffin, J. and Rajasingham, L. (1995), *In Search of the Virtual Class: Education in an Information Society*, London: Routledge.

Wilss, L. (1997), *Computer Assisted Learning at Queensland University of Technology: Student Learning Process and Outcomes*, Brisbane: QUT.

Update – Supporting organizational change: fostering a more flexible approach to course delivery

From 1999, the significance of online teaching increased. It was viewed as a powerful example of a flexible delivery strategy. To overcome some of the confusion and miscommunication associated with the earlier flexible delivery policy a series of round-table discussions, chaired by the Deputy Vice Chancellor and Pro-Vice Chancellor (Information and Academic Services), were organized involving key stakeholders across the University. Their involvement indicated the commitment of support of the senior executive of the University.

Through the round-table discussion a Framework for Online Teaching was adopted which communicated an important message. For the foreseeable future Queensland University of Technology would remain a campus-based university with a strong emphasis on face-to-face teaching. However, online teaching would be used across the board to enhance student access to information and to facilitate communication with teachers and peers. An increasing proportion of teaching staff would also complement their traditional teaching with a range of online strategies such as discussion groups, model answers to exam questions, access to online resources and multiple-choice questions to test student understanding of key concepts. A small number of courses, primarily in the field of Information Technology at postgraduate level, would use online delivery as the teaching approach of choice.

This framework established a common vocabulary for online development. At the beginning of 1999 three separate online systems were emerging. Education and Information Technology each had a separate system and Information and Academic Services was developing an in-house system that was being used in a limited manner across all faculties. Given concerns about the sustainability of separate systems, a central initiative grant of $650,000 was made available to pool resources and talent and contribute to a single online system. The grant was allocated to the Co-ordinated Online Teaching (COLT) consortium. In 2000, a teaching and learning innovation fund of $1 million was established to build on the work of the COLT group and encourage the strategic adoption of online teaching strategies in all faculties. Faculties were encouraged to bid for up to $100,000 to adopt online teaching strategies progressively, which was allocated on an incentive rather than competitive basis.

These measures have led to a more focused and strategic approach to the integration of technology in the actual processes of teaching and learning, and in embedding technological literacy as one of the key attributes of QUT graduates.

Contact author

Professor Gail Hart has fifteen years' experience in the HE sector in Australia. She is currently the Director of the Teaching and Learning Support Services (TALSS) Department at Queensland University of Technology. She was a member of the national Committee for University Teaching and Staff Development (CUTSD) from 1997 to 1999. Her professional expertise includes mental health nursing, organizational development and change and human resource management. [*g.hart@qut.edu.au*]

Staff development at RMIT: bottom-up work serviced by top-down investment and policy

Carmel McNaught and Paul Kennedy
Royal Melbourne Institute of Technology, Australia

Initially published in 2000

Effective staff development is the weaving together of many strands. We need to support staff in their current work, while providing them with ideas, incentives and resources to look for new ways to design learning environments which will enhance student learning. Staff development must be combined with specific projects where change is occurring. Ideas are not hard to find. Incentives and resources are another matter. The paper will outline some general principles for effective staff development. These principles will be applied in the description of the substantial investment RMIT has made in order to realize our teaching and learning policy. We have a model of 'grass-roots' faculty-based work funded by large-scale corporate 'investment'. 'Bottom-up' meets 'top-down'.

Educational design as the key to successful flexible learning

What is the business of a university in the 1990s and 2000s? Quantity and quality are both important considerations in modern universities as they seek to maintain important intellectual and physical spaces for their staff to pursue creative research and development, while at the same time needing to provide teaching for escalating numbers of students in all courses in order to shore up funding. These student cohorts have become increasingly diverse (McInnis *et al*, 1995) with more part-time students and students from a greater variety of backgrounds. Flexible modes of delivery have been widely viewed as the prime way of meeting the challenges posed by this diversity. There has been a fair amount of naïve equating of flexible delivery with production of online materials ('Plug them into the Web') and insufficient attention to the relationship between flexible modes of operation for students, the use of communication and information technologies, and the design of educationally sound learning environments (Kennedy and McNaught, 1997; Reeves and Reeves, 1997). However, there is no doubt that communication and information technologies will be a major part of future university planning, as several recent reports make clear (e.g. Yetton *et al*, 1997).

Discussions about using technology for flexible learning often centre on variation in time and place access to learning experiences. But it must mean more, if we are to believe that technology can meaningfully enhance students' learning experiences. Looking at how we can cater for a variety of learning styles, for example, by offering a variety of learning activities and a variety of assessment strategies, is essential.

Good educational design is the key to successful flexible learning. Here at RMIT University we offer staff a set of online tools to assist them in refurbishing their subjects and courses. We explain the functionality of each of the tools in terms of student learning activities. Table 1 matches some student learning needs, with examples of the design of suitable student activities, with components of the online toolset. Several of the tools could be used for most of the activities; examples are used for simplicity.

Student learning need	Examples of student activity	Example of current RMIT benchmark Distributed Learning System (DLS) toolset component
Information handling skills	Web searching; using electronic Library databases	
Developing understanding	Building links between information from various sources; problem-solving exercises	*CourseInfo/ BSCW* *Question Mark*
Linking theory to practice	Working with embedded media and simulations in course material; tutorial programs with feedback	*CourseInfo* Hybrid systems with CD-Roms
Practising discussion and argument	Online debates using a threaded discussion	*WebBoard*
Practising articulation of ideas	Role playing using a threaded discussion; sharing essays online	*WebBoard* *BSCW*
Rehearsing skills and procedures	Online quizzes with feedback	*Perception Question Mark*
Practising teamwork	Group projects	*BSCW*
Learning professional practice	All of the above!	

(The student learning need is based on Laurillard: *www2.open.ac.uk/LTTO/internal/tsaa.htm*)

Table 1: Functions of the RMIT DLS toolset

Universities as organizations which support or hinder innovation

In a recent investigation into the factors supporting the adoption of computer-facilitated learning (CFL) at Australian universities (McNaught *et al*, forthcoming), three major themes emerged. These were *policy*, *culture* and *support*. The considerable overlap between and within these themes is illustrated in Figure 1. There needs to be a congruence of policy, culture and support factors if significant adoption of CFL strategies is to occur.

The policy theme looked at specific institutional policies, such as equity and intellectual property, the alignment of policy throughout the organization, the direction of policy change (bottom-up or top-down) and a number of strategic processes which flowed on from policies such as grant schemes. Culture incorporated factors such as collaboration within institutions, and personal motivation of staff to use CFL, as well as particular aspects of funding, staff rewards and time, leadership, teaching and learning models, and attitudes such as 'not invented here'. Support incorporated a whole gamut of institutional issues including IT, library and administrative infrastructure, professional development for staff, student support, educational and instructional design support for academic staff, funding and grant schemes, and IT literacy.

Several universal factors in relation to widespread use of CFL were identified:

- coherence of policy across all levels of institutional operations and specific policies which impact on CFL within each institution;

- intellectual property, particularly the role of copyright in emerging online environments;

- leadership and institutional culture;

- staff issues and attitudes, namely, professional development and training, staff recognition and rewards, and motivation for individuals to use CFL;

- specific resourcing issues related to funding for maintenance or updating of CFL materials and approaches, staff time release and support staff.

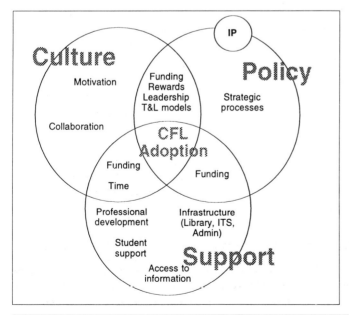

Figure 1: Themes and their relationships affecting the adoption of CFL

Staff development and training

In all universities this is seen as a vitally important area. We should not underestimate the difficulties involved in innovation and change. Marris (1974) parallels the sense of loss during bereavement to the resistance one can feel when letting go of known ways of doing things and embarking on new strategies. For many academics the increasing emphasis on the use of computer technology for administration, research and teaching is highly threatening. We need to recognize these fears and devise plans which build staff confidence and motivation, and provide adequate support and training opportunities.

Staff development can no longer be a pleasant 'cottage industry' on the fringes of academe or the enthusiastic enterprise of a few individuals supported by 'soft' money. Effective staff development is positioned at the centre of university functioning and yet needs to retain connections with the needs and perceptions of teaching staff. This is a demanding challenge. Staff development programmes that are successful in meeting the needs of complex modern Australian universities need to be supported strategically (and financially) by their own universities.

Hughes *et al* (1997), in a study of twenty Australian universities, describe three approaches to staff development for the use of information technology in teaching – integrated, parallel and distributed. These approaches are defined and the discussion in Hughes et al. is summarized in Table 2. In reality, universities use a combination of approaches, though with a trend in one direction. The table is useful as a tool for assessing the potential strengths and weaknesses of the combination of any particular set of support units in a given university

The number of players in the professional development area is large, including:

- 'Traditional' academic development units, concentrating on general teaching and learning support; these can be centrally located or in faculties.

- Units where the key focus is the use of communication and information technologies in teaching and learning. These can be centrally located or in faculties. They are often called flexible learning units.

- Units which focus on courseware production using technology. These can be centrally located or in faculties. Some of these units have evolved from print-based distance education units or centrally-based Information Technology Services units.

- University libraries.

Ellis *et al* (1998) reported on an online survey of twenty academic development units (48 per cent response rate with a follow-up phone survey conducted of non-respondents) about staff development activities for technology in teaching and learning undertaken during 1997 and those planned for 1998. Results show that most of this type of staff development is still delivered by traditional methods such as classroom presentations, demonstrations and half-day tutorials while online methods of delivering staff development are less frequently used. The content of these courses covers a broad range of topics with the most popular being pedagogical issues in online course design, Web page design, and course authoring systems. Staff undertaking training tended to be from a cross-section of academic levels. Staff development activities of this nature are not

Integrated Approach (eggs in one basket!)
Strong structural links between units or section of the one unit which provide general T&L support, support for using IT in T&L, and production support for courseware. Essentially top-down.

Benefits:	*Issues raised by:*
Coherent policy framework.	Ease of access by all staff limited.
Efficient planning of resources and	Individual approaches less likely to be
avoidance of duplication.	recognized.
	An emphasis on one technological solution
	may emerge and overwhelm educational
	design.

Parallel approach (never the twain shall meet?)
Separate units for general T&L support and support for using IT in T&L

Benefits:	*Issues raised by:*
Allows due recognition to be given to a	Co-operation between the various units may be
wide range of T&L issues (e.g internationalization) and	difficult to achieve. There is a potential for
not just educational design associated with the use of IT.	confusion and competition to emerge.
Allows the development of expertise relating to the	May result in a narrow range of educational issues
new technologies.	being addressed in the IT in T&L units.

Distributed approach (organic sprouting)
More bottom-up than the other two approaches. A range of units, centrally located and in faculties which are not tightly co-ordinated. Project management remains with local projects.

Benefits:	*Issues raised by:*
An 'organic' solution where unnecessary controls	Can result in weak project management where
do not hamper innovation.	there may be insufficient educational expertise.
Can be economical as skills are sought when they	Potential for innovations to falter without visible
are needed.	institutional support.
	Can result in waste and duplication of effort and
	resources, including equipment.

Table 2: Integrated, parallel and distributed approaches to staff development for the use of information technology in teaching (after Hughes et al., 1997)

exclusively provided by the academic staff development unit, but tend to be carried out by a range of internal and external providers, as noted above. The academic development units often play a key role in establishing and maintaining relationships between these units and the co-ordination of their activities is essential to the development of coherent and comprehensive staff development programmes.

McNaught *et al* (forthcoming), identify six key issues in staff development:

• The appropriate balance point between centrally provided and local staff development services needs to be determined in each university. Central services can be more clearly linked to university priorities; faculty or department services can be more in touch with local needs.

- As technology becomes more mainstream, support services need to be scaled up. This involves deciding on the level of support that can be afforded and the model of support which is most apposite. The educational design and evaluation, technical, and media production support services that universities currently have are under strain. It is unlikely that the existing examples of good practice at each university will be sufficient to ensure that new or revised subjects will be well designed and evaluated. By modelling good practice themselves, mentors can assist staff to make optimal use of resources.

- A follow-on issue is determining the optimal relationship between staff development and production support services. Again, this needs to be decided in each university context.

- Even if an integrated model of professional development is adopted, there are still many professional development providers at most universities. Mapping the services of each provider and ensuring reasonable co-ordination is increasingly important as the need for support services scales up.

- Academic and general staff work load is a key issue. Careful work planning to ensure that staff have time to learn new skills and manage new processes is essential.

- We are in a time of rapid change. It is important that professional development support be flexible, appropriate and adaptable. It should make sense to staff, be linked to practice and be appropriately timed.

Just how effective are academic development unit activities in supporting adoption of CFL?

Figure 2 shows how staff in academic development units (ADUs) at Australian universities rate various activities in terms of how effective they believe each activity is in increasing the uptake of CFL in their university. The data in the survey in McNaught *et al* (forthcoming) were represented by a five-point scale from 'very important' (5) through to 'not important' (1); the data have been collapsed into two categories – 'important' (4 and 5) and 'limited importance' (1–3) in order to see trends more clearly.

This can be compared with the responses of seventy-three members of ASCILITE (Australasian Society of Computers in Learning in Tertiary Education – the Australasian equivalent of ALT) – who rated how effective they believed each activity had been in supporting their use of CFL materials (Figure 3). The profiles are remarkably similar. Educational design and individual consultations are believed to be most important. However, as many ASCILITE members are innovators or early adopters (see Figure 4 below), this congruence between the perceptions of providers and clients must be tempered with the need to provide staff development across the whole range of staff expertise and interest. Indeed, coverage of support for all staff, not just the enthusiast teachers, has always been a major issue for academic development work.

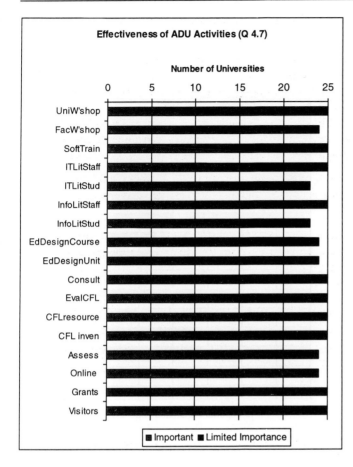

Figure 2: ADU Assessment of effectiveness of various activities in supporting the uptake of CFL

Key:
UniW'shop: General workshops across the university
FacW'shop: Faculty/ department workshops
SoftTrain: Software training sessions
ITLitStaff: IT literacy support for staff
ITLitStud: IT literacy support for students
EdDesignCourse: Educational design of entire courses
EdDesignUnit: Educational design of units
Consult: Individual consultations
EvalCFL: Evaluation of computer-facilitated learning (CFL) innovations
CFLresource: Providing information about CFL resources
CFLinven: Maintaining an inventory of CFL projects in the university
Assess: Support for computer-based assessment systems
Online: Support for online learning system
Grants: Facilitation of grant writing for CFL development
Visitors: Visiting specialists, teachers, scholars

Are all staff being supported?

A recent survey of ASCILITE members (McNaught *et al*, forthcoming) showed interesting data about the perceptions innovators or early adopters have about their colleagues. Most of the seventy-three members surveyed regarded themselves as innovators or early adopters (Figure 4) and many had developed significant projects single-handedly with little support from faculties or their university. These members were able to see the need for a well-supported environment for development. They were asked to categorize themselves on the scale:

- innovators;
- early adopters;
- users when technology is mainstream;
- very reluctant users.

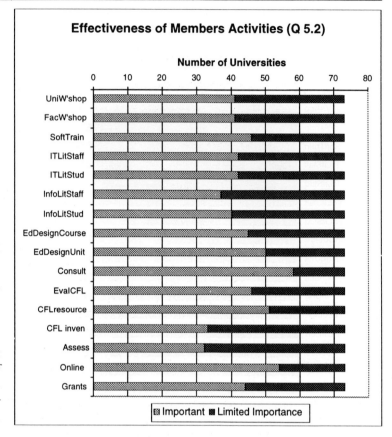

Key: See Figure 2. Note that ASCILITE members were not asked about the effectiveness of visiting speakers and scholars as several members work in this capacity themselves.

Figure 3: ASCILITE members' assessment of effectiveness of various activities in supporting their uptake of CFL

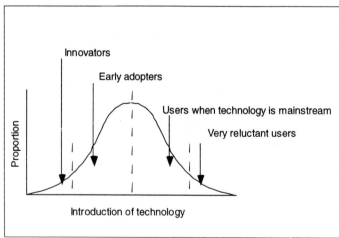

Figure 4: Schematic diagram of phases of technology take-up (after Rogers, 1995)

We also asked them to consider where the majority of staff in each category of department/ faculty/ university were on this scale. The results are shown in Figure 5. The data from the survey were in four categories; the data have been collapsed into two

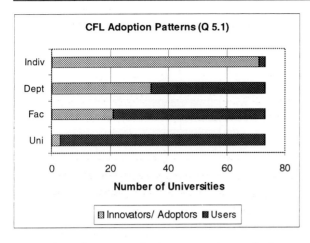

Figure 5: ASCILITE members' perceptions of CFL adoption patterns at their universities

categories – innovators/ adopters and users/ reluctant users – in order to see trends more clearly. It is striking how isolated in many ways these innovators/early adopters are. The majority of respondent ASCILITE members considered themselves to be innovators or early adopters while they perceived that the majority of staff at their institutions only used technology when it was mainstream or were very reluctant users.

How big are the staff development needs?

Data obtained from Information Technology Services units at Australian universities provide insight into another issue relating to staff using computer-facilitated learning strategies. Figure 6 illustrates software support available to staff and their use of it. Note that while the respondent universities were all able to provide data about university infrastructure software support, many did not comment on staff and student use. It is clear that staff do not use the full range of technologies available to them. There are complex issues relating to culture, staff development and adequate provision of facilities at a local level that relate to the fact that the majority of Australian academics use their computers for email, Web-browsing and maybe basic Web teaching, and office applications (*Word*, *PowerPoint* and *Excel*). Also, it may be that some technologies will not be considered appropriate by the majority of staff and will not be used widely. Of course, it is heartening to note that many universities have set in place useful infrastructure for software support but, as indicated above, staff will use technology in their teaching when culture, policy and support structures are congruent.

Applying these ideas to the context of RMIT University

Universities in Australia are currently in an environment of intense change. They are being required to educate more students, from an increasing variety of backgrounds, with decreasing government funding. Universities are required to compete vigorously for student enrolments and external sources of funding. In this environment, universities have had to reassess their fundamental business and the way they go about it. Information technology (IT) is viewed as an important factor in streamlining their operations.

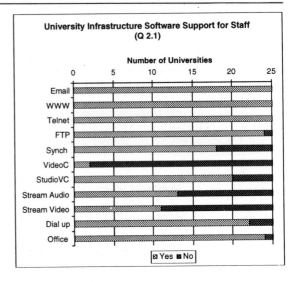

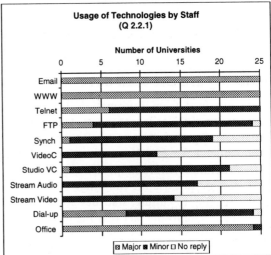

Figure 6: Software support available to staff and their usage

Key: email, Web, telnet, ftp, synchronous chat, desktop videoconferencing, studio videoconferencing, streaming audio, streaming video, dial-up access, general Office applications

RMIT University is an 'old' (in Australian terms; RMIT began in 1887) technological university. It is highly diverse – it is a bi-sectoral (includes vocational sector) university and has the largest number of international students of any Australian university. There are seven strong faculties which often resist central directions. In recent years there has not been a strong staff development programme.

In the program which is described below, RMIT wanted staff development which:

- is linked to RMIT business and vision;

- promotes sound educational practice;

- ensures flexible learning is 'owned' in every department;

- organizes adequate support for all staff;

- results in low increase in staff work loads.

There are two key policy documents which are currently guiding the direction RMIT takes for the next three to five years. The first is the Teaching and Learning Strategy (T&LS).

RMIT teaching and learning strategy

The RMIT teaching and learning strategy aims to provide a student-centred learning environment where:

- subject programmes and the courses they comprise are designed to develop the following graduate attributes in students: knowledgeable, critical, responsible, creative and with a capacity for life-long learning, leadership and employability and an international outlook;

- the system is flexible enough to suit the particular learning needs of students in terms of their prior experience and current situation;

- courses are designed and implemented holistically with coherent connections between subjects comprising the core of a course;

- students and the community are seen as significant stakeholders;

- assessment is directly related to the explicitly stated objectives of subjects;

- quality improvement and quality assurance based on reflective practice and customer-focused systems design are ubiquitous.

There are resources allocated to implement the T&LS both in human and financial terms. For example, each faculty has two senior positions established by secondment of academic staff members from within the faculty. Each faculty has a developing Faculty Education Services Group (FESG) where technical and educational support for staff is available.

RMIT IT Alignment Programme

RMIT University established a project team in 1998 to develop an Information Technology Strategy designed to facilitate the implementation of the objectives of the Teaching and Learning Strategy in respect of electronically mediated flexible learning environments. The RMIT Education and Training Alignment Project (ITAP) report (1998) delivered by the team in June 1998 and adopted by the University, forms the basis for a $A50 million investment by RMIT over the next three years (1999–2001).

The report comprises several elements:

- IT infrastructure aligned with the needs of education to deliver the systems and hardware necessary to provide students with an electronically connected learning environment and access to computer-based learning resources;

- a Distributed Learning System (DLS) compliant with the emerging Educom/CAUSE Instructional Management System (IMS);

- a Student Management System (SMS), fully integrated with the DLS to provide enrolment and subject and course progress records electronically accessible to academics and students;

- an extensive review of all academic processes within the university in a Business Process Re-engineering (BPR) project;

- extensive staff development.

Enacting the RMIT teaching and learning strategy through the IT alignment project: designing the RMIT distributed learning system

We have to deliver on our promise that we can provide a flexible set of tools that will enable staff who are not technological whizz kids to develop pedagogically sound, interesting, and relevant online courses in an efficient and well administered way. How have we designed our Distributed Learning System (DLS)? Here are some of our principles:

- a suite of tools, not just one;

- integrating educational principles into the description of the toolset;

- IMS compliance of all tools;

- a team approach to all online projects;

- involvement of all seven faculties in a benchmarking exercise to evaluate the toolset and the effectiveness of the learning environments we are building.

A learning-centred evaluation is being attempted. In order to set up a base-line for the teachers' reflections, teachers in each DLS project are asked to articulate the student learning outcomes for their subject and where they think the online experience would enhance learning. We ask all teachers to submit a weekly journal entry via an online feedback form to continue this process of reflection. It is quite difficult to extract this stream of continuous reflection from teachers (staff workload remains an issue), but several teachers do give us formal and informal feedback from time to time. As much of the informal anecdotal feedback is in email messages, the substance can be captured.

We also use the usual evaluation strategies with students of online questionnaires, focus groups, analysis of Web access data, analysis of support/ help desk records, and analysis of performance on learning outcomes. At this stage we really have only anecdotal evidence of learning enhancement (or otherwise).

Each semester we use a spreadsheet to collate the feedback data collected from staff and data received from staff and students from the DLS Help Desk. This is supplemented by data from student surveys conducted by individual subjects and reports from focus groups with students. The data are then coded using an iterative method whereby each discrete statement is assigned a descriptive category; these are then reviewed with some reduction of the number of categories occurring. For example, in Semester 1 of 1999 this process led to the identification of thirty-nine categories under which the comments were grouped. The categories were grouped under seven headings – access, toolset competence, support, student issues, educational outcomes, communication and miscellaneous. A report on the evaluation of Semester 1 1999 DLS subjects is given in McNaught *et al* (1999).

Staff development through the Learning Technology Mentor Programme

Approximately seventy Learning Technology Mentors have been appointed – one in each department of the university and some in central areas such as the Library. These are mostly academic and teaching staff who have funded one day a week time release to develop online materials and support their colleagues in their departments to engage with online teaching and learning. Initially this is a one-semester program but it will be extended in 2000, giving some LTMs a further 26 days' time release, and bringing another cohort of 120 LTMs on board.

These LTMs undertake an extensive staff development programme about a week long. Some of the key topics are:

- RMIT's vision with respect to the university's position as a major international technological university. The Boyer (1990) Scholarship model has been used for some time as an integrating model for all RMIT work.

- The evolution of the Teaching and Learning Strategy over the last few years.

- The structure and function of the ITAP; description and key staffing of the ITAP Teams. Some comment on the importance of the Business Process Re-engineering (BPR).

- Course and subject renewal guidelines exist in all faculties and form a central focus of the T&LS and the way in which ITAP works. The concept of graduate attributes is part of this process.

- Roles of the faculty-based Faculty Education Services Groups (FESGs). Relationship between FESGs and central ITAP Teams.

- Overview of the DLS toolset; how the use of the DLS tools relates to the renewal of subjects.

Additional staff development sessions are run each week. These sessions cover a range of practical 'hands-on' sessions and workshops in areas such as assessment and evaluation strategies for online learning, student induction methods, managing digital resources, project management, etc.

All LTMs develop a work contract with the first author who heads the Professional Development Team of the ITAP; if individual staff wish this can be formalized into accreditation for a subject in a Graduate Certificate of Flexible Learning.

We are seeing that staff development and support for developing online learning materials and strategies must become distributed across the organization. Therefore the role of the faculty-based Faculty Education Services Groups (FESGs) is pivotal. Growth needs to occur in these units rather than at the centre. We believe that technical support staff, educational designers and graphical designers are needed at faculty level and the only courseware production that should exist at the centre is some support for high-end media production and multimedia production. We are trying to combine the benefits of both the integrated and distributed approached mentioned earlier by Hughes *et al* (1997).

Where to from here?

We have a great deal of consolidation and development to do. We have been delighted by the enthusiasm of many Learning Technology Mentors. We have a sense of gathering momentum. In one year we have 190 subjects using the Distributed Learning System and many more in planning for use in 2000. Several faculties are showing real commitment, though a couple might still need a persuasive nudge. Have we reached critical mass yet, where the appropriate use of technology will roll out across the University? Probably not, but we feel we are on the right track.

References

Boyer, E. L. (1990), *Scholarship Reconsidered. Priorities of the Professoriate*, Princeton, New Jersey: The Carnegie Foundation for the Advancement of Teaching.

Ellis, A., O'Reilly, M. and Debreceny, R. (1998), 'Staff development responses to the demand for online teaching and learning', in Corderoy, R. (ed.), *FlexibilITy: The Next Wave?* Proceedings of the Australian Society for Computers in Learning in Tertiary Education Conference, University of Wollongong, 191–201.

Hughes, C., Hewson, L. and Nightingale, P. (1997), 'Developing new roles and skills', in Yetton, P. (ed.), *Managing the Introduction of Technology in the Delivery and Administration of Higher Education*, Evaluations and Investigations Program report 97/3, Canberra: Australian Government Publishing Service, 49–79, *http://www.detya.gov.au/highered/eippubs1997.htm.*

Kennedy, D. K. and McNaught, C. (1997), 'Design elements for interactive multimedia', *Australian Journal of Educational Technology*, 13 (1), 1–22.

Laurillard, D., 'Technology strategy for academic advantage', Open University, UK, *http://www2.open.ac.uk/LTTO/internal/tsaa.htm.*

Marris, P. (1974), *Loss and Change*, London: Routledge & Kogan Page.

McInnis, C., James, R. and McNaught, C. (1995), *First Year on Campus*, a commissioned project for the Committee for the Advancement of University Teaching. Canberra: Australian Government Publishing Service.

McNaught, C., Kenny, J., Kennedy, P. and Lord, R. (1999), 'Developing and evaluating a university-wide online Distributed Learning System: the experience at RMIT University', *Educational Technology and Society*, 2 (4), *http://ifets.gmd.de/periodical/vol_4_99/mcnaught.html.*

McNaught, C., Phillips, P., Rossiter, D., and Winn, J. (forthcoming), *Developing a Framework for a Usable and Useful Inventory of Computer-facilitated Learning and Support Materials in Australian Universities*, Evaluations and Investigations Program report. Canberra: Higher Education Division Department of Employment, Education, Training and Youth Affairs.

Reeves, T. C. and Reeves, P. M. (1997), 'Effective dimensions of interactive learning on the World Wide Web', in B. H. Khan (ed.), *Web-based Instruction*, Englewood Cliffs, New Jersey: Educational Technology Publications, 59–66.

RMIT Education and Training IT Alignment Project,
http://www.online.rmit.edu.au/main.cfm?code=ia00.

Rogers, E. M. (1995), *Diffusion of Innovations*, Fourth Edition, New York: The Free Press.

Yetton, P. and associates. (1997), *Managing the Introduction of Technology in the Delivery and Administration of Higher Education*, Evaluations and Investigations Program report 97/3. Canberra: Higher Education Division Department of Employment, Education, Training and Youth Affairs,
http://www.deet.gov.au/divisions/hed/operations/eip9703/front.htm.

Contact author

Carmel McNaught is the Head of Professional Development in Learning Technology Services at RMIT University. She integrates her staff development programme with working as an educational designer and evaluator in specific projects where staff are using communication and information technologies. Another field of research involves looking at relationships between academics' beliefs about teaching and learning and their use of computers for these purposes. [*carmel.mcnaught@rmit.edu.au*]

Learning technology in a networked infrastructure

Improving instructional effectiveness
with computer-mediated communication

Som Naidu, John Barrett and Peter Olsen
University of Southern Queensland, Australia

Initially published in 1995

This study explores the use of asynchronous Computer-Mediated Communication (CMC) in the delivery of instructional content, and points up the interaction among learners, as well as between learners and instructors. The instructional content in the project described was available to learners online as Microsoft Word documents, with email being used for communicating within the student group. Many students, as well as some of the instructors, felt uncomfortable with the flexibility and openness that a CMC environment allowed. However, once familiar with this process of instruction and interaction, learners were able to work consistently at their own pace, and understand that instructors are interested in every individual learner's opinion and in the collective views of the group. It was evident that a CMC-based instructional delivery system, when carefully planned, has the potential to facilitate that outcome, and to improve instructional effectiveness.

Introduction

A critical question that often faces educational technologists is how to deliver excellence in teaching and subject-matter content to learners. A corollary to this question is how and what instructional technologies can be brought to address this search for particular contexts? While there is a wide range of instructional delivery technologies we can choose from, our choice must be carefully considered. The number of factors to be considered are too many to list here, but must include a cognizance of the nature of the content or skill that comprises the subject matter of instruction, the learners, the time and their place of study, and the costs of the delivery mode, both for the learners and the institution. This paper reports our experience of the first phase of a three-phased integration of Computer-Mediated Communication (CMC) aimed at improving instructional effectiveness.

Interest in the application of some form of CMC in the enhancement of teaching-learning environments is currently widespread. This has led to the emergence of a growing body of literature on various aspects of CMC-based instructional delivery systems (for example, Mason and Kaye, 1989; Harasim, 1993; Mason, 1993; Wells, 1993). While there exists in this literature a great deal of information on the hardware and software requirements for such delivery systems, and numerous reports of applications in a wide variety of contexts, relatively little attention is focused on approaches to the integration of CMC in teaching-learning environments.

The project reported in this paper is about the design of instructionally effective CMC-based teaching-learning environments. The emphasis here is not on the hardware or the software requirements but on the organization and presentation of subject matter, the designed activities, and human factors which are often, in our experience, at the heart of the success or failure of such projects. A phased-integration of computer-mediated communications technology is described and also recommended to overcome an all-too-pervasive mindset about teaching that is a carry-over from the conventional face-to-face classroom instructional situation.

Teaching-learning orientation

A considerable amount of research exists in favour of teaching-learning designs that engender collaboration and interaction among the peer group (Ide *et al*, 1981). There is evidence, for example, that co-operative learning benefits learning for all except the most concrete, repetitive tasks with effect sizes as high as $.80\sigma$ (Johnson and Johnson, 1974). Allowing learners to exercise adaptive control over their learning process is also reported as having positive impacts on learning (Hannafin and Colamaio, 1988). CMC-based teaching and learning systems are characteristically designed to empower learners by allowing them a greater degree of adaptive control over their learning environment which is not feasible in conventional systems. This facility refers to the flexibility that learners may have in exercising as much control and autonomy over their learning, as and when necessary.

The hypothesized advantages of this electronic teaching-learning environment over the conventional system were several. For the learners it was intended to encourage them to move from:

- a defined learning space to an open, richer global resource base;

- an instructor-controlled learning environment (that of a conventional classroom) to a collaborative and co-operative learning context; and

- being uncritical recipients of content to dynamic and participant explorers of the knowledge base.

The orientation of this project is derived from increasing evidence in favour of CMC-based collaborative learning and instructional environments. Its basic philosophy of flexible access and an instructor-learner negotiated teaching and learning process is at the heart of open learning systems, and increasingly now of conventional face-to-face teaching-learning environments which are also aiming for more flexible and adaptive formats (see the concept illustrated in Figure 1).

Figure 1 is a graphic representation of the teaching-learning environment that was designed and implemented as part of this study. Notice that the focus in the design is on integrating CMC in the teaching-learning environment as one of the resources available to the teacher and the learner. Students enter this instructional environment with variable

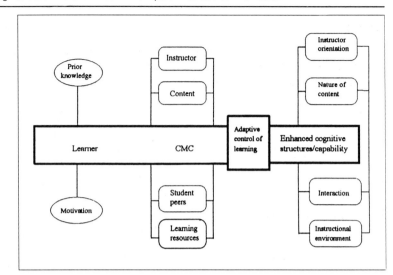

*Figure 1:
Integrating CMC
in the teaching-
learning process*

levels of prior knowledge in the use of CMC as a teaching and learning delivery medium, and also with variable levels of motivation to use it. Upon entry, they are confronted with an instructional environment that comprises the instructor, subject-matter content and CMC-based learning activities, conventional group lecture sessions, tutorial assistance, and other forms of individualized and group support. The goal of this instructional scenario is to allow the learner a level of *flexibility* that will enable *adaptive control* of the instructional environment. For the learner, this means seeking out and receiving as much instructional direction as is necessary, or assuming as much flexibility and freedom as one desires. Such autonomy should also allow the learners to use the instructional resources in a manner that will suit not only their place and pace of study, but also their learning styles and approaches to learning.

The intended outcomes of this instructional environment are several. These include foremost, for the learner, changed cognitive structures. This means a dramatically new and improved way of relating to content, storing it in memory and retrieving it. For the instructor, this arrangement implies new and improved ways of delivering content, and also improved ways of interacting with learners. The instructional delivery system also facilitates the arrangement of subject-matter content in new and interesting dimensions. Content can be organized in hypertext formats and include hypermedia to facilitate a richer and resourceful instructional and learning environment. All of these outcomes in the end lead to the creation of a new and improved instructional system that is more flexible, richer, and a realistic teaching and learning environment.

Context and subjects

The project reported in this paper was carried out in the context of a second-year unit in the Bachelor of Education program in the Faculty of Education at the University of Southern Queensland (Australia). The unit titled *Teaching-Learning Studies* focused on the classroom teacher and related issues like planning for teaching, instructional strategies, and classroom management. The project ran for a full 13 weeks in Semester 2, 1994. Subjects in the study were all the students enrolled in the unit.

Design

In order to ensure systematic integration of CMC in the teaching and learning context, and to cope with the human factors pervasive in the situation, a three-phased model for implementing CMC was devised (see Figure 2).

A *Document source*	B *Communications*
3 Integrated MM	3 Desktop video + audiographics
2 Hypertext documents	2 Computer conference (groupware)
1 Microsoft Word	1 Microsoft Mail

Figure 2: The three-phased CMC implementation

There are two components to each phase. The first component (A) concerns the organization and presentation of the subject matter. At the first level, all subject-matter content can be created, stored and retrieved as word-processed files. At the second level, the word-processed files are presented as hypertext documents to facilitate such uses as complete document search, a linked hypertext-type lesson structure, zooming-in and magnifying components, copy and paste to other documents, and insert 'post-it' notes and 'bookmarks', etc. The third level extends the hypertext document to include interactive multimedia and activate other programs such as a statistical package.

The second component of the system, the communications dimension (B), is similarly organized into three layers with the first level operating on electronic mail. The second level includes text-based conferencing systems and/or groupware, while the third level includes audiographics and/or desktop video conferencing.

This system provides for a 'stepwise' progression allowing advancement up the layer on one dimension while operating at a lower level in the other dimension (that is, hypertext documents and email – A2 B1), or integrated multimedia documents and conferencing – A3 B2). The building of these steps is dependent on the technical infrastructure, skill level of students and the teachers, and more importantly, on the nature of subject-matter content and the desired interactions/activities between and among all participants.

Figure 3 represents the implementation model reported in this paper. The current project was designed to represent the first phase of this model. Core content in the model is held on a file server at level A1 as *Microsoft Word* documents. Level A2 uses CD-ROM as a storage device while level A3 utilizes greater mass storage devices such as *Interleaf* and *Worldview*. Communication at level 1 was via *Microsoft Mail*. At level 2, computer conferencing is envisaged, and at level 3 we are looking at audiographic communication.

A Document source	B Communications
3 I-CON Author + Interleaf + Worldview	3 Smart 2000
2 Interleaf + Worldview	2 Computer conference
1 Microsoft Word	1 Microsoft Mail

Figure 3: USQ implementation model

Network configuration

The phase 1 network configurations included 486 33 EISA machines running *Windows 3.0* with *Microsoft Word 2.0* and *Microsoft Mail*. The core lecture content, including the learning activities for each week's study, was input as Word documents. All supporting reading materials were scanned and also included as Word documents. Students were required to log on, read each week's material, and then respond to the required learning activities via *Microsoft Mail*. Students in the project were registered on the local-area network individually, as well as under an alias which was the unit number. A user could therefore send a message to another directly by addressing a message to that individual, and also to the whole group by addressing messages to the alias.

Learning activities

Learning activities were generated by the instructor on a weekly basis. These activities were carefully designed to focus on the subject-matter content that was being covered during that week. In the first week, for example, students examined the topic of 'direct instruction' as an instructional strategy. Students were exposed to a discussion on the topic by the instructor. This was accessible as a Word document. After reading and synthesizing this material, students were required to respond to the learning activities that were generated by the instructor. Some activities required further reading, others required some field work, and others required summarizing or synthesizing.

On-line response patterns

Responses to these instructor-generated learning activities were transmitted via email to the student-group alias, which meant that it went out to all the students registered under that alias, including the instructor. Each student's response was therefore received by all the others in the group. Individuals could comment on all the responses that were made, and especially those that raised controversial or interesting issues. The instructor was in a position to observe the transactions that were going on via email and make his own contributions to the discussion. He would make his own contributions and also provide feedback to the student commentaries. Students could also send messages to individuals in the group, which would then comprise private communications as neither the instructor nor anyone else in the group could have access to that message, unless such a message was copied or forwarded to a third or fourth individual in the group.

Data sources

The objective of data collection was focused on the use and utility of the CMC-based delivery environment to the stakeholders (that is, instructors and learners). Hence mostly ethnographic data-gathering techniques were employed. These included in-depth interviewing of learners during and after the process, and also analysis of the texts from the online interactions. Questions asked, focused on various aspects of the delivery system and their interface with it. These included training in the use of email; the benefits and disadvantages of the delivery mode; the changed role of the instructor; the technology-human interface in online communication; impacts on learning outcomes; and problems with reading text online and other concerns relating to online communication. All interview protocols were tape-recorded and transcribed for further analysis.

Analysis of Data

The interviews were conducted using the following specific headings:

- training for the project;
- instructional delivery system;
- changed role of the instructor;
- technology-human interface;
- learning experiences.

Observations

Preparatory training

Learners reported being generally satisfied with the training that was provided in the use of email and online communication at the start of the project. There was a view that the training could have been more effective if it had been focused on specific needs of learners with variable levels of expertise with the email software and online communication. Most felt that a list (in print form) of the most common features of Microsoft Mail would have helped a lot. Some of the learners found certain components of the training redundant, and which could therefore have been eliminated with better design, delivery and co-ordination of the training programme.

The instructional delivery system

Participants felt that, while this new mode of delivery was 'different', it offered them much needed flexibility in the way of their place, pace and time of study. The general disposition was that this was a more time-efficient way of making content and instructor expertise more accessible to learners.

Changed role of the instructor

The use of CMC in instructional delivery meant a significant change in the role of the human teacher. While CMC allowed the instructor a greater degree of interaction with individual students in the class, it also increased the instructor's workload. Even though learners realized that they were in contact with their instructor online 24 hours a day, the lack of face-to-face contact with the instructor in a CMC-based instructional environment was missed by the students. There was an overall preference for retaining some face-to-face contact in this predominantly electronic teaching-learning environment.

Technology-human interface

For the computer-literate, a CMC-based teaching-learning environment posed little problem. For the computer-illiterate however, this delivery mode was a source of much anxiety. The experience from this project revealed that several factors, if not carefully managed, could lead to the failure of such systems. These include issues relating to moderation of discussions, careful management of dominant personalities, and use of acceptable online conversation protocols. The asynchronous nature of the medium meant that a response from the intended recipients of a comment was not available or 'visible' immediately as is the case in face-to-face contexts. A few of the participants found it rather frustrating having to wait for 24 or more hours to get some response on their comments and queries.

Some students found reading and composing on-screen somewhat difficult. This was not unexpected. Initially, students preferred to read offline, compose their message on a word processor, then send the message as an attachment to an email message. This was a very time-consuming and tedious exercise that detracted from one of the obvious advantages of CMC systems, which is the spontaneity with which messages can be received and responded to. With some practice, however, students were able to read comfortably on-screen and also compose messages online.

Learning experiences

A majority of the students felt that one of the significant outcomes of the CMC-based delivery system was the shift in the degree of control learners could exercise in their learning. In the conventional face-to-face context, their instructors made a lot of the decisions about how and when things were supposed to happen. In the CMC-based system, the learners took over a greater degree of responsibility of their learning patterns, determining for themselves how and when things would be done. In some cases there was a commensurate increase in the amount of time students spent on studying. Students enjoyed the opportunity to interact with their peers in an on-going manner without being constrained by the hours of a lecture and tutorial session. Students reported getting more 'involved' in their study.

There were some negatives as well. One of the problems that seemed to concern students about online communication was that of being misunderstood and not able to defend or explain one's comments before the others were already in an attack mode. In a face-to-face mode, if one felt that a comment was being misunderstood or misrepresented, there was the opportunity to correct it right there and then. In a CMC-based delivery mode, however, once a message had been posted, it was out in the open for its recipients to make whatever out of it, and it was only after some time, usually after a day or so, that one could correct oneself or clarify the misunderstanding of the others. An advantage of the possibility of this occurring was that learners were forced to think through their ideas and comments a lot more clearly before broadcasting these online. Some students found that in so doing they were improving their reading, writing and thinking skills.

Discussion

A CMC-based delivery system is fundamentally different from a face-to-face teaching-learning environment. It carries with it, among other things, a uniquely different student and instructor mindset. Like any other instructional delivery system, a CMC-based teaching-learning environment has its strengths and weaknesses for learning and teaching outcomes. Its success, in these terms, is dependent on a number of design considerations that are unique to it. These considerations and their implications on teaching and learning in a CMC-based instructional delivery system are discussed in the remainder of this paper.

Implications for system management

An integral part of the success of a CMC-based instructional system is the function of the system manager. This is a person who is responsible for network upkeep and maintenance. Teaching and learning using CMC requires maintenance and support of email communications, word processing and transferring of files and material from word-processed documents to mail and vice versa. Many things do and can go wrong in this process, especially when dealing with students and instructors who lack much experience with operating in this mode.

It must be understood by all parties concerned that there is a protocol for online communication. These have been widely documented in the literature on CMC-based instructional systems (see Harasim, 1993). Stakeholders must be given initial training in these and their use with continuous support and reinforcement throughout the process. It must be understood that there will be a slow start, and that as students and instructors become familiar with the delivery medium, the pace will pick up. This underscores the need for that front-end training and on-going support to be carefully planned and executed.

Implications for the instructor

This refers to the perceptions that instructors hold about their roles as teachers. Usually these perceptions are a result of their own training as teachers, or requirements of their institution, and years of work experience in conventional or other educational systems. The principal tasks of teachers in conventional face-to-face instructional systems, for instance, are to prepare lectures, deliver them, hold small group tutorial sessions, and mark written assignments.

A CMC-based teaching-learning environment prescribes a different set of tasks for the instructor. These include, among other things, the development of the core content in advance of the actual instruction, presenting these online, and communicating with students on a continual basis in an asynchronous mode. The instructor takes on a facilitator's role, reading and responding to student's work and clarifying any misconceptions as they arise. The instructor is no longer tied down to lecture or tutorial times. All of this happens asynchronously and in a 'fluid' teaching-learning context. For some instructors this may mean additional work which may lead to the fear of being eternally 'on the job'. The truth is that a CMC-based instructional environment, if carefully planned and executed, has the potential to save time and effort for the instructor – time which previously had been taken up by routine and repetitive tasks such as grading paper-based work and answering generic types of queries.

Many instructors also find this to be a rather 'loose' arrangement in which they relinquish their control over how students study. This is not always acceptable to many instructors who believe in retaining greater control of their classes. In the academic community, some of these perceptions are deeply ingrained and very resistant to change. Much of this resistance, especially towards new delivery technologies, can also be attributed to a lack of confidence on the part of instructors in their abilities, and a fear of having to learn a new skill. A CMC-based instructional delivery system poses this kind of threat to most instructors, especially those who have little or no proficiency in working with students online.

Our experience tells us that these are serious considerations that must be carefully planned for. Instructors have to be introduced to the radical shift in the mindset or orientation towards their teaching function in a CMC-based environment. Unless this shift in their mindset has taken place, success is far from certain.

Implications for the learner
Students, also, hold particular conceptions and a mindset about their own roles as learners and that of their instructors. Like instructors, they need to understand that a CMC-based teaching-learning environment shifts the bulk of the responsibility for learning onto themselves. This requires recognition of that responsibility on their part, and the assumption of an active role in the learning process. Students must understand that the instructor is no longer going to drive the learning for them, that they must learn to drive it for themselves.

Importance of front-end training
While an increasing number of the current generation of students are computer- and Internet-literate, there are still many among them who have no or only negligible experience with computers and electronic networks. Experience derived from this project suggests that a carefully planned front-end training is imperative for both students and instructors involved in any CMC-based teaching-learning arrangement. Such a training program must be based on a thorough assessment of the needs of the participants. It is likely that these needs will be variable which would then necessitate individualized or small-group attention for particular or all aspects of the training. Yet while front-end training is a must, it is also certain that training up-front in the use of CMC will be insufficient. Training and continuing assistance will be necessary throughout the duration of such a project. Therefore, help with online instruction and the equipment will need to be available on a continual basis, either online or otherwise.

Outcomes of CMC-based instruction

A CMC-based instructional environment is fundamentally different from the conventional classroom-based system in very many ways. The difference between the two modes of instruction is most explicit in the manner in which content is delivered to the learner, the communication patterns between tutors and students and among the students themselves, the manner in which students access additional learning resources, and the manner in which assessment of learning is carried out. Most instructors who have been exposed to a CMC-based instructional system argue, although not initially, that the mode has had a significant and lasting impact on their overall approach to teaching and learning. Our experience suggests that this kind of an impact is manifested in several ways. Some of the more visible ones are discussed in the following sections of the paper.

Expanded resource base

Currently, the integration of CMC in instructional systems may include any one or more of a growing number of applications. These include access to electronic mail, bulletin boards, databases, networked CD-ROMs, network newsgroups, electronic discussion lists and computer conferences, File Transfer Protocol (FTP) facilities, and global search tools such as Gopher, Telnet, Knowboots, Netfind, Finger, and Archie. This list tends to be expanding all the time. With the help of an average personal computer and a communication connection, these applications have the potential of bringing to the learners an expanded learning-resource base which would be otherwise beyond their reach. The learning resources that these applications can bring to the learners include, *inter alia*, libraries and other sources of information at remote sites, access to international experts and research sites, other students, colleagues and peers, and access to relevant discussions that might be taking place in other parts of the world.

Our experience and those of others (Ferris and Roberts, 1994) show that access to an expanded learning resource base that is carefully selected can mean that learners are not only encountering *more* useful material than that specified in the curriculum, they are also covering the specified content and more material more rapidly than is possible in conventional systems. Learners shift from being passive receptacles to being active participants in the search for knowledge. Moreover, they learn how to acquire and use knowledge, and instructor-roles shift from dispensers of information to producers of environments which allow learners to learn as much as is possible on given topics.

Co-operative learning environment

Conventionally, co-operative learning refers to instructional and learning environments which are characterized by increased interactions among individual learners or among small groups of learners. This form of learning has its limitations no doubt, but its positive contributions to learning and instructional outcomes are reported to far outweigh its limitations or difficulties (Johnson and Johnson, 1984). A computer-mediated learning and teaching environment which depends upon interaction between databases, instructors, tutors and students, and also among students, is by definition a co-operative one. Computer-mediated co-operative learning and teaching has been shown to enhance learning outcomes in many different ways, including improvement in the quantity and quality of the learning experience (see Collis, in press).

Computer-supported co-operative instruction is also reported to have benefited the instructors. Ferris and Roberts (1994) report that as a result of the experience in the use of CMC in their instructional environments, instructors began to take a larger and more systems-oriented perspective of their contributions in the school, meeting each week in teams to plan an integrated effort. Problems that arose were dealt with by the teams rather than by the one instructor in the conventional system. Almost subconsciously, the roles of instructors in their classrooms got transformed into a more collaborative one.

Authentic learning environment

A computer-mediated learning and instructional environment is capable of reaching out to resources in remote locations, including people and sites, for relevant information that would otherwise be impossible to access by most students in the time that is usually available to them. As such, a CMC-based instructional system is capable of facilitating 'authentic' learning by enabling access to environments separated in time and place from one's home base. Realities can be created, in cyberspace in a manner of speaking, by electronic access to people *in situ* in different countries, cultures, systems and with access to libraries, databases and also discussions raging on the Internet.

Authentic assessment environment

Similarly, a CMC-based instructional environment is potentially capable of an assessment system that is authentic and dynamic in nature. Assessment can be authentic in the sense that it is situated and contextualized rather than contrived. It can be dynamic in that assessment is continuous and pegged to what students are doing in terms of their interactions and activities online. A variety of instructional strategies are open to instructors in CMC-based instructional systems such as setting up discussions, debates, tasks, project work etc. to which students are required to contribute. These contributions are asynchronous, and can be assessed by the instructor on an individual basis, including the provision of individualized as well as group feedback.

Flexible learning

Flexibility is a characteristic feature of a CMC-based learning and instructional environment. The integration of some form of CMC in teaching and/or learning means that the communication channels between the instructor and the students, and among the students, are not only open all the time, but *asynchronous*. With communication channels *open*, teachers and students have access to one another at any time of the day rather than waiting for the lecture, tutorial or consultation times to raise a query or share some interesting thought as these are occurring in the process of one's study. Messages, questions and contributions by students or teachers can be left on the network asynchronously, meaning at different times and from different locations. Open and *asynchronous* communications between the instructor and the students are the hallmarks of a flexible learning arrangement.

A CMC-based learning and instructional system is open also in terms of the many instructional strategies it can accommodate. These include individual searches of online databases, journals, libraries, and discussion groups. One-to-one communication may take the form of correspondence study, learning contracts, apprenticeships, and internships between the instructor-student, the student-expert, and also between students. One-to-many online communication may include lectures, symposiums, and panel discussions. Many-to-many techniques can include debates, simulations/games, role plays, case studies, discussions, project-based work, brainstorming, delphi and nominal group techniques, forums, and cognitive networking/mapping. This list of online activities is by no means exhaustive. It is an indication of what is possible in a CMC-based instructional environment with a little bit of imagination and creativity, mostly as part of instructional design and development.

Learning skills development

A CMC-based learning environment places the responsibility for learning more than ever in the hands of the learner. In so doing, the learner is greatly empowered, and also placed in a position of greater control of not only the amount but the quality of his or her learning. This autonomy allows learners (especially the enterprising ones) to explore, experiment, take risks and venture beyond that which is necessary. A natural outcome of this kind of initiative on the part of learners is enhanced learning skills relating to the search for and acquisition of knowledge. In conventional systems of instruction, much of this kind of autonomy is not possible due to a greater degree of instructor-control of learning, and also because of the lack of resources in the learners' immediate learning environment. With the help of CMC, that learning environment is now much larger and more accessible, allowing for a richer and larger resource base for the learners as well as the instructors.

Concluding comment

This project was implemented to ascertain, *inter alia*, factors that needed addressing when building a CMC-based learning and instructional delivery system. It was evident, from our experience, that in the initial stages students as well as instructors require an intensive training program to familiarize them with basic operations such as logging on and the use of passwords, moving between applications, such as from Word to Mail, and how to save and store material for future reference. Many students and instructors are still locked into the time and place concepts associated with lectures and tutorials. The issues of flexibility and openness must be carefully introduced to all stakeholders in such a project. Once familiar with this process of instruction and interaction, learners are able to work consistently at their own pace, and realize that instructors and tutors are interested in every individual learner's opinion and also in the collective views of the group. It is evident that a CMC-based instructional delivery system has the potential to facilitate that outcome as well as improve instructional effectiveness.

Acknowledgement

The authors are grateful to the Committee for Advancement of University Teaching (CAUT) for financial support in the conduct of this study. In addition, we thank John Elms and Ken Woolford (Faculty of Education, University of Southern Queensland) for their support.

References

Collis, B. (in press), 'Cooperative learning and CSCW: research perspectives for internetworked educational environments', in Lewis, R. (ed.), *Lessons from Learning*, Amsterdam: North Holland.

Ferris, A. and Roberts, N. (1994), 'Teachers as technology leaders: five case studies', *Educational Technology Review*, 3, 11–18.

Hannafin, M. J. and Colamaio, M. (1988), 'The effects of variation of lesson control and practice on learning from interactive video', *Educational Communication and Technology Journal*, 35 (4), 203–12.

Harasim, L. M. (1993), *Global Networks: Computers and International Communication*, Cambridge MA: MIT Press.

Ide, J. K., Parkeson, J. A., Haertel, D. D. and Walberg, H. J. (1981), 'Peer group influence on educational outcomes: a quantitative synthesis', *Journal of Educational Psychology*, 73 (4), 472–84.

Johnson, D. W. and Johnson, R. T. (1984), 'Cooperative small-group learning', *Curriculum Report*, 14, 1–6.

Johnson, D. W. and Johnson, R. T. (1974), 'Instructional goal structure: cooperative, competitive or individualistic', *Review of Educational Research*, 44 (2), 153–66.

Mason, R. (1993), *Computer Conferencing: The Last Word*, BC Canada: Beach Holme Publishers.

Mason, R. and Kaye, A. (1989) (eds), *Mindweave: Communication, Computers and Distance Education*, Oxford: Pergamon Press.

Wells, R. (1993), *Computer-mediated Communication for Distance Education: An International Review of Design, Teaching, and Institutional Issues*, Research Monograph No. 6, Pennsylvania State University College of Education PA, ACSDE.

Update – Improving instructional effectiveness with computer-mediated communication

The development of computer-supported collaborative learning and teaching environments has been used to maximize teaching and learning outcomes in teacher education programmes offered by the University of Southern Queensland. Naidu and Olsen (1996) analysed a third-year education unit which utilized Computer-Mediated Communication (CMC) to promote interaction and reflection regarding issues such as planning for teaching, instructional strategies and classroom management. The analysis revealed that this mode of instruction fostered asynchronous and synchronous interactions between individual students and among groups of students. We were interested in how the use of CMC could help learners to become reflective thinkers, refine their teaching skills in the practice environment, and build networks with peers, mentors and university staff.

Following the success of this model, Naidu and Olsen attempted to find out if CMC could be used to develop reflective analysis of teaching practice episodes. The same unit was utilized to carry out this project. Each week students were given questions to assist them to reflect about aspects of teaching and how these related to their own experiences in the practice teaching environment. These reflections were emailed to tutors and other group members for comments and discussion. Students and tutors became actively involved in commenting, questioning and analysing these reflections. The last three weeks of the course comprised an online conference relating to the analysis of Glasser's classroom management techniques.

The results of this project varied from group to group. One important factor to come out of the project was that the role of the tutor was crucial for the success of this particular teaching/learning environment. One tutor had problems with the technology and another had trouble in allocating time to respond and interact with the students. Apart from these incidents, the results of student surveys and interviews confirmed that this method suited their learning and development in gaining more insight into effective teaching and learning episodes in practice teaching situations.

Reference

Naidu, S. and Olsen, P. (1996), 'Making the most of Practical Experience in Teacher Education with computer supported collaborative learning', *International Journal of Educational Telecommunications*, 2 (4), 265–78.

Contact author

Dr Som Naidu is now Associate Professor in the Multimedia Education Unit, and adjunct Associate Professor in the Department of Learning and Educational Development, Faculty of Education, University of Melbourne.
[*s.naidu@meu.unimelb.edu.au*]

From the sage on the stage to what exactly? Description and the place of the moderator in co-operative and collaborative learning

Chris Jones
CSALT, Lancaster University

Initially published in 1999

This paper reports a significant finding from a two-year study of computer conferencing used to deliver a course unit at a UK university. Computer conferencing has been applied to education alongside a concern to develop co-operative and collaborative learning strategies. The technology of computer conferencing has been identified as especially appropriate to a co-operative style of work. This study found that far from collaboration being an outcome of the deployment of computer conferencing it became in some sense the problem. A common 'gloss' on the educational changes that are taking place, with the introduction of new technologies for teaching and learning, is that the 'sage on the stage' is being replaced by 'the guide on the side'. This paper argues that this opposition rests on little substantial evidence or research. The moderator/facilitator role advocated as suitable for computer conferencing is shown to be deeply embedded in wider social actions. The orientations of the tutor are heavily inclined towards the demands of assessment. Successful computer-supported collaborative learning (CSCL) is the outcome of the co-operative work of all the members of the conference. The application of CSCL relies upon timely interventions by the tutor.

A collaborative technology?

In a standard text on open and distance learning, computer conferencing appears with advantages and disadvantages that rely on the capacity of the conference to 'facilitate' students and staff working together (Mason, 1994). Co-operation or collaboration are seen as unproblematic outcomes of the use of the computer conference. Hiltz noted that collaboration is both technically facilitated and becomes a moral requirement that involves a respecification of the teacher's role (Hiltz and Benbunan-Fich, 1997, p. 1). The outlook that associates collaborative learning with computer conferencing can be found throughout a wide range of literature (Mason and Kaye, 1989; Harasim, 1990; Kaye, 1992; McConnell, 1994; Harasim *et al*, 1995; O'Malley, 1995; Jonassen, 1996).

In some educational literature co-operation and collaboration are treated as distinct approaches (Topping, 1992). In this paper they are treated as interchangeable, an approach justified by the indistinguishable definitions provided by many authors – see for example McConnell (1994); O'Malley (1995). Co-operative or collaborative learning is understood to mean learning that takes place with students and tutors working together on academic tasks. It is argued to be fundamentally different from the direct transfer of knowledge by a largely one-way transmission, the 'transmissive' model (Harasim, 1990; Hiltz and Benbunan-Fich, 1997). The terms are glossed over so as to provide an image of a movement from the 'sage on the stage' to the 'guide on the side'. It can only be noted in passing that the transmissive model contains many elements of co-operation and collaboration (Macbeth, 1990).

Moderator

Two recent texts illustrate the new role promoted for teachers associated with CMCs and computer conferencing. Jonassen comments:

> Your role as the teacher must change from purveyor of knowledge to instigator, promoter, coach, helper, model and guide of knowledge construction. (Jonassen, 1996, p. 261)

Mason draws a comparison with other media:

> The role of the computer conferencing teacher is the farthest removed from that of the traditional lecturer . . . During the course . . . the teacher's role is definitely one of facilitator and host rather than one of content provider and star of the show. (Mason, 1994, p. 42)

The two accounts converge on a style of teaching that emphasizes social skills and reduces the emphasis on content and personal delivery. The delineation of the moderator/facilitator's function has been a recurrent feature of writing and research on computer conferencing (Kerr, 1986; Feenberg, 1989; Mason, 1991). Mason comments that:

> The idea of the expert teacher must give way to a network of supports and resources in which everyone has some kind of expertise to be tapped. (Mason, 1994, p. 47)

This paper explores empirically the practice of a tutor orientated toward providing a moderating role with a view to examining the claim that computer conferencing leads to the adoption of a facilitator/moderator role.

The background to the research

The research was undertaken at Manchester Metropolitan University between 1994 and 1996. The course unit, 'Technology in Communications', was part of the BA in Information Technology and Society Degree (BAITS). The unit was taught online using the *FirstClass* computer conferencing system.

> Within the constraints imposed by the availability of technology, the communications technologies under investigation will be used to deliver the course. Material will be sent either to student's own machines or to machines made available in the lab. This 'online' approach will be supported by face to face meetings. (BA Information Technology in Society Student Handbook 1994 supplement)

The online nature of the course was accompanied by an expectation that work would be done co-operatively in small groups without lectures.

Methodology

An ethnographic methodology was employed to generate an adequate description of 'just what' happened when university education was transposed from a traditional setting into the new technology. Within education the qualitative approach to research is well represented, in particular the ethnographic tradition (Hammersley, 1986 a,b; Fetterman, 1984, 1986; Eastmond, 1995). The aim is to understand the setting from the point of view of those involved in it. Ethnography is an intrinsically descriptive task that resists formalization. The methods used are 'naturalistic' in the sense that they rely on the study of people and their activities in their natural environment (Fetterman, 1998). The form of ethnography adopted has been heavily influenced by work in CSCW (Computer Supported Co-operative Work) and ethnomethodology (Suchman, 1987; Heath and Luff, 1992; Hughes *et al*, 1993).

Findings

This paper will examine the work undertaken by the tutor who was responsible for both the course design and delivery. Within the conferencing system the tutor had the 'Moderator' role. The moderator in *FirstClass* is the person in charge of a specific conference, determining who can contribute and setting levels of access, in contrast to the 'Administrator' who has control of the conferencing system as a whole. The educational role of the tutor, in organizing the course, determining the content and arranging assessment, was combined with an organizational role built into the software.

Course design

Prior to the course, the tutor had identified a potential problem of the course having 'weaker' structures online.

> Students are given responsibility to do the work, what happens if they don't work? – therefore deadlines – therefore a schedule of tasks, readings for example, log on, send a message etc. (Interview course tutor 10/94)

Continuous assessment and deadlines set throughout the course were devised as prompts to encourage the students both to work and to work online. A surprising omission from the original course documentation was any developed idea of collaborative working. In the first weeks of the course students were given handouts explaining course features. These notes implied a form of group collaboration but there was little explicit acknowledgement of this element. The course documentation placed little emphasis on collaborative and co-operative working. The emphasis was on the 'flexibility' available for students in terms of when work might be done, and the opportunity for synthesis and reflection. In contrast the expectation of collaboration was clearly expressed at the first face-to-face meeting.

The course outline provided the mechanism for an anticipatory account of the course. It was through the production of this documentation that the tutor was accountable to the students and wider institution for the content and teaching methods of the course. It is through the detailed control of what work is to be done and how that work is done that the tutor uses the resource he/ she has provided in the course documentation and planning. The following two examples explore how the collaborative nature of the course emerged and how the assessment procedure developed.

Collaboration

... well let me just emphasise a number of points, computer conferencing is a collaborative activity you have to work together, you don't just go off and work alone; you have to work together through the conferencing system. (Tutor introducing course unit 10/94)

The tutor made extensive comments on the process of the work done and gave advice for future work. The first part of one message is included in Table 1.

Friday, November 3, 1995 10:59:27 am
TIC95 MODULE 1 Item
From: J. A. C
Subject: Pats on the back, advice and a bit of a warning
To: TIC95 MODULE 1
1   I have been very pleased to see that a significant section
2   of TIC-95 students are acquiring the on line skills which
3   we have been discussing over the last few weeks. I have
4   received a number of replies via my post box to the
5   queries that I raised on the work for module one.
6   Furthermore, in at least two of the sub-conferences we
7   have seen real debate about the constitution of groups
8   and how marks should be allocated. ....
15   .... Some of the messages I have
16   received have avoided the issue of relative
17   contributions. I don't want to make a big issue out of this
18   but I do want to award marks fairly so that effort is
19   rewarded.

Table 1: An example of tutor advice and comments

The tutor clearly puts a great emphasis on the process of collaboration and use of the system (lines 1–8). Equally emphasis is placed upon allocating marks to individuals so that 'effort' is rewarded (lines 18–19). This concentration on how the work was done inverted the intention to stress the content of the course not the 'system' or the method of work. It would seem that the day-to-day concerns of the tutor outweighed the original design consideration that the content of the course should not be submerged by the mechanism of delivery. It was seen by the tutor as a diversion from the work he intended.

Assessment

The question of individual marks illustrates a further tension within the structure of the course. Working together and online activity were stressed by the tutor, but marks were allocated individually. The tutor wanted students to collaborate and provide information in a way that students believed would undermine group coherence. Within many educational environments marks are allocated individually and students are graded separately, so there is a tension between the pressure of assessment and the demands of group work.

The tutor's comments throughout the course show him to be concerned with academic standards, a defender of course quality. Sections of an extensive message to all students are reproduced in Table 2.

5	First and foremost the mark for a particular group has been
6	determined by the quality of the piece submitted. I have looked
7	for originality, evidence of research into the topic and the
8	degree to which the question set has been answered.
9	Secondly I have looked at the way in which group members
10	have worked together in constructing the answer and the
11	efficiency with which material coming from different individuals
12	has been put together into a coherent whole.
	(Tutor to Module 2 23/1/96)

Table 2: Academic standards

The tutor began by setting out standard marking criteria. Later a plea was made for analysis 'which goes beyond description' dealing with abstract and conceptual issues. The tutor was advising students on how to do academic work and also how to talk about it. The tutor ends by advising students to read two sub-conferences as examples of good practice (Table 3).

40	As you move to the end of this unit and prepare for your third year,
41	you should be beginning to produce analytical work which goes
42	beyond description and shows that you can deal with conceptual
43	and abstract issues. We can learn from one another in this unit as
44	all the work is available on the conferencing system. Please have
45	a look at other contributions and see if you can find useful ideas.
I	
46	recommend everyone to look at the report produced in the BLL
47	group. This was a model of how material should be organised and
48	how a coherent answer can be constructed. I recommend that you
49	look in the xxxxx conference to see a good example of how a group
50	sought to work together and had a good deal of success and some
51	failures in achieving cooperation.
	(Tutor to Module 2 23/1/96)

Table 3: Tutor direction and advice

There were two related standards being applied, a common academic standard of intellectual discourse and a more specific injunction as to how to work within the conferencing system. This displays an orientation in the conference through which the tutor is encouraging the social process of doing conferencing as well as providing academic guidance. The names of the two sub-conferences were garbled in the original message, possibly as the message was translated between word-processing packages. No one notified the tutor about this. I can only assume that no student followed the academic advice and tried to learn from the model sub-conferences. The request for students to 'learn from one another' fell on deaf ears.

An exceptional case
There were several students who found group work difficult. One of these students, Ben, produced work alone, which presented the tutor with a difficult assessment problem. In November 1995 the tutor sent a message clearly setting out his requirements:

No work will be accepted from individuals in module 2 and this means that anyone not joining a group is effectively opting out of this course unit. (Tutor 7/11/95)

Despite these clear guidelines he received a submission from a lone student and faced a number of students who were clearly not co-operating.

Now my dilemma is what I should do about this . . . I want them to work together. The other problem is the anomalies, Ben's a complete anomaly . . . Ben's produced a good piece of work but he cannot work with others, so what sort of moderation can you bring in about being unable to work with others, . . . (Field Notes 16/1/96)

Ben was clearly thought by the tutor to be of an exemplary academic or intellectual level, his submission was identified as a model showing the required academic standard.

From 12 February Ben sent one message a week asking for collaborators and suggesting possible topics. On 18 March Ben announced;

I plan to work alone on a project as it looks as if there is little interest in collaboration from others in the unit. I have not finalised the subject area as yet . . . (Ben to TIC95 Projects 18/3/95)

This message was followed by another in Projects on 22 March. It was not until 16 April that Ben asked for a sub-conference to work in. 'Roadrunner' was set up reflecting a declared interest in the link between road protesters and new technology. No messages were posted to the conference apart from the course tutor until 2 May 1996:

no road project following a problem finding a focus . . . attached is my digital audio broadcasting project submission. (Ben to 'Roadrunner' 2/5/96)

The tutor had begun with a rule that groups should have a maximum of five and a minimum of three members. At the end of the unit he had a conference with a lone member who had not posted to the conference and had changed his subject without notification.

On 7 May the tutor commented:

Then there is the problem of Ben. Last time I looked there was only my letter in the conference and he has changed his subject at the last minute; there is a real problem here. Do I mark him for non-co-operation? (Field Notes 7/5/96)

When the marking had been completed I discussed assessment again and was told:

Ben's was the best but he didn't collaborate. I've penalized him for working on his own. If I was blunt I would say it was mainly his fault . . . effectively he refused to collaborate. So his work is reduced from 70 to 65 per cent, because a percentage of the mark was for collaboration the mark for the project was less than the content. Let me be clear it was a notional reduction. I mean, how do you decide on the reduction? (Field Notes 18/6/96)

The tutor had posted marks to all the sub-conferences explaining the marking scheme for the project and a separate message reminding groups of the overall weighting that gave projects 50 per cent of the course mark (Table 4).

Within the marks massaging the 'notional' reduction for Ben takes on a formal appearance because a definite proportion of the marks, 30 per cent, is identified as being awarded for the way the work was produced.

```
Tuesday, June 4, 1996 2:30:28 pm
xxxx Marks Item
From:      John A. Cawood
Subject:   Project Mark
To:        xxxx Marks

In marking these projects, I have taken into account firstly the content and originality of the report itself
and secondly the use that the project team made of the conferencing system and how well individuals
worked together. I have awarded 70% of the marks to the report itself and 30% to the process of its pro-
duction.
```

Table 4: Project mark

Ben's case casts light on the way in which definite course aims are subject to the working out of the contingencies of the course's development. The 'plan' for all students to be in groups is repeatedly unrealized and the tutor has to make *ad hoc* day-to-day decisions about whether to allow the situation to continue and then how to assess a student that is so clearly in breach of expressed course criteria.

Discussion and conclusions

The nature of collaborative effort was orientated towards the expectations implicitly and explicitly embedded in the course structure. The original course design mentioned little about collaboration but was clearly designed around groups achieving the completion of a series of modules. The tutor's initial comments raised the profile of collaborative working as a key component of the course. How these features of the course were achieved was then subject to *ad hoc* and mundane interventions. Collaboration was not something that could be pre-planned, either by good software design or by the pedagogic design of the course unit. It relied upon *in situ* day-to-day management of the conference.

This conclusion echoes some debate within the CSCW community about the status of plans (Bardram, 1997). Collaboration as planned action has an ironic status, in that the members of the conference first must work together, collaborating in order to produce the context for their joint activity. The scheme for collaboration is a resource for the work rather than in any real sense determining its course. The assessment criteria clearly did not provide clear and definite instructions for dealing with Ben's case. A range of contingent and situational factors was brought to bear, including the student's wider status as a 'good' student. The tutor had to decide if and how he could penalize *this student* for his failure to co-operate. The practical implication of this observation is that the timely interventions of the tutor were a significant feature enabling the plan of collaboration and accredited academic success to be achieved.

The institutional position of the tutor is exhibited in the production and use of the course aims and objectives. The tutor controls the definition of what counts as success within the conference. Assessment confirms that the tutor is in the position of holding specialist and superior knowledge. Despite the equation of computer conferencing with collaborative learning and the shift from 'knowledge-giver' to facilitator, the wider educational context remains one of assessment and accreditation. The tutor is the first line of that institutional system of accreditation of knowledge, determining what counts and how much is enough.

There is of course a second sense in which the tutor remains in command of superior knowledge despite no longer 'delivering' lectures. The tutors are likely to know the answers to the questions they have set. Despite the emphasis on students as a resource, and on collaborative and group learning techniques, the tutor has a responsibility to 'know' the subject area. Even when relying on the students' own experience, the arbiter of the validity of these experiences remains the tutor, who is assumed to have a superior subject knowledge to the student. The tutor thus remains in command of superior knowledge despite no longer 'delivering' lectures.

The moderator's role was superficially a more social and organizational role than that provided for in the 'transmissive' approach. However, it is clear that the tutor is orientated towards marking and assessment criteria that are deeply embedded in the wider institutional arrangements of the university. The 'technology of accountability' within the educational setting ensures that the tutor remains in possession of superior knowledge (Suchman, 1993). In so far as the tutor becomes a guide it is still unclear whether they have to remain with the sage up on the stage. The tutor had hoped that the computer conferencing system would assist him in being able both to give a group mark and judge individual contributions. This was never achieved but the tutor used his plan to accomplish the marking consistently in the form of an individual allocation of marks. The plan for assessment was unwaveringly inadequate to guide the tutor without respecification of its terms to deal with the variety of ways in which students orientated themselves to the course objectives. In this way the plan had a reciprocal relationship with activity; the plan was itself made intelligible by the actions and in turn made those actions intelligible.

1. Collaboration was a *moral* imperative not technologically determined in any strong or weak sense.

2. The role of the tutor as moderator could be said to be *constitutive* of collaboration rather than an outcome of the pedagogy and technology.

3. The observations show how the *ad hoc* and day-to-day interventions of the tutor are orientated around the requirements of assessment.

Acknowledgement

The research for this article was conducted at the Department of Information and Communications and the Department of Sociology at Manchester Metropolitan University, United Kingdom.

References

Bardram, J. E. (1997), 'Plans as situated action: an activity theory approach to workflow systems', in Hughes, J. A., Prinz, W., Rodden, T., and Kjeld, S. (eds.), *ESCW'97: Proceedings of the Fifth European Conference on Computer Supported Cooperative Work*, Dordrecht: Kluwer Academic Publishing.

Eastmond, D. V. (1995), *Alone but Together: Adult Distance Study through Computer Conferencing*, Cresskill, NJ: Hampton Press, Inc.

Feenberg, A. (1989), 'The written world: on the theory and practice of computer conferencing', in Mason, R. and Kaye, A. (eds.), *Mindweave: Communication Computers and Distance Education*, Oxford: Pergamon.

Fetterman, D. M. (1984), *Ethnography in Educational Evaluation*, Beverly Hills, CA: Sage.

Fetterman, D. M., and Pitman, M. A. (eds.) (1986), *Educational Evaluation: Ethnography in Theory, Practice and Politics*, Beverly Hills, CA: Sage.

Fetterman, D. M. (1998), *Ethnography Step by Step* (2nd edn), Applied Social Research Methods Series Volume 17, Newbury Park: Sage.

Hammersley, M. (ed.) (1986a), *Controversies in Classroom Research*, Milton Keynes: Open University Press.

Hammersley, M. (ed.) (1986b), *Case Studies in Classroom Research*, Milton Keynes: Open University Press.

Harasim, L. (ed.) (1990), *Online Education: Perspectives on a New Environment*, New York: Praeger.

Harasim, L., Hiltz, S. R., Teles, L. and Turoff, M. (1995), *Learning Networks: A Field Guide to Teaching and Learning Online*, Cambridge, MA: MIT Press.

Heath, C. C., and Luff, P. (1992), 'Collaboration and control: crisis management and multimedia technology in London underground line control rooms', *CSCW Journal*, 1 (1–2), 69–94.

Hiltz, S. R., and Benbunan-Fich, R. (1997), 'Supporting collaborative learning in asynchronous learning networks', *UNESCO/Open University International Colloquium: Virtual Learning Environments and the Role of the Teacher*, Milton Keynes, Open University, *http://eies.njit. edu/~roxanne/*.

Hughes, J. A., Somerville, I., Bentley, R. and Randall, D. (1993), 'Designing with ethnography: making work visible', *Interacting with Computers*, 5 (2), 239–53.

Jonassen, D. H. (1996), *Computers in the Classroom: Mindtools for Critical Thinking*, Englewood Cliffs, NJ: Merrill, Prentice Hall.

Kaye, A. R. (ed.) (1992), 'Collaborative learning through computer conferencing: the Najdeen Papers', *NATO ASI Series F*, Berlin: Springer-Verlag.

Kerr, E. (1986), 'Electronic leadership: a guide to moderating on-line conferences', *IEEE Transactions on Professional Communications*, PC, 29 (1), 12–18.

Macbeth, D. H. (1990), 'Classroom order as practical action: the making and un-making of a quiet reproach', *British Journal of Sociology of Education*, 11 (2), 189–214.

Mason, R. (1991), 'Moderating educational conferencing', DEOSNEWS, 1(1).

Mason, R. (1994), *Using Communications Media in Open and Flexible Learning*, London: Kogan Page.

Mason, R., and Kaye, A. (ed.) (1989), *Mindweave: Communication, Computers and Distance Education*, Oxford: Pergamon.

McConnell, D. (1994), *Implementing Computer Supported Cooperative Learning*, London: Kogan Page.

O'Malley, C. (ed.) (1995), *Computer Supported Collaborative Learning*, Berlin: Springer-Verlag.

Suchman, L. (1987), *Plans and Situated Actions: The Problem of Human-machine Communication*, Cambridge: Cambridge University Press.

Suchman, L. (1993), 'Technologies of accountability: on lizards and airplanes', in G. Button (ed.), *Technology in Working Order*, London: Routledge.

Topping, K. (1992), 'Cooperative learning and peer tutoring: an overview', *The Psychologist: Bulletin of the British Psychological Society*, 5, 151–61.

Update – From the sage on the stage to what exactly? Description and the place of the moderator in co-operative and collaborative learning

Since I wrote this article describing the actual work done by an online tutor, I have continued to carry out research investigating both tutors' and students' experiences of networked learning. Practitioners moderating and developing courses in networked learning environments still have difficulty anticipating the outcomes of their designs (Jones *et al*, 2000). While practitioners of networked learning have adopted a new paradigm at the level of theory for teaching and learning, there remains a gap at the level of 'rules of thumb', a common sense or common knowledge that might be applied to the tutor's role in networked learning. There are some signs that this lack of research-based practice, developed from the practice of teachers and learners, is beginning to change.

Other authors have produced significant additions to our knowledge of the role of educational practitioners in these new educational environments (Salmon, 2000). There is beginning to emerge a picture of practice that can be generalized and codified for both newcomers and seasoned practitioners. This is a positive gain, and we should welcome the emphasis on moderation as a skill that requires development. It remains a fundamental insight that the technology does not determine collaboration or entail any particular model of teaching or learning. The skilful day-to-day activity of a tutor is required to ensure the success of networked learning.

The place of accreditation in the role which the tutor has to play is also being more fully recognized (Brown and Duguid, 2000). The moderator's role includes 'credentialing', providing the detailed marking and assessment criteria upon which the university rests. It remains a distinguishing feature of universities, distinct from other learning organizations, that they provide accepted evidence of learning. Brown and Duguid call this function 'warranting'. They point out that warranting entails representing learning to the individual and making a declaration about the knowledgeable status of individuals to society. Warranting and credentialing remain key parts of the tutor's role.

References

Brown, J. S., and Duguid, P. (2000), *The Social Life of Information*, Boston MA: Harvard Business School Press.

Jones, C., Asensio, M. and Goodyear, P. (2000), 'Networked learning in higher education: practitioners' perspectives', *ALT-J*, 8 (2), 18–28.

Salmon, G. (2000), *E-moderating: The Key to Teaching and Learning Online*, London: Kogan Page.

Contact author

Chris Jones is a research lecturer in the Department of Educational Research at Lancaster University and currently teaches on the MSc / Diploma in Advanced Learning Technology and the online course Networked Open Learning. He is conducting research as part of the JISC/CALT-funded project Networked Learning in Higher Education. He has until recently taught a variety of courses in politics and public administration.
[*c.r.jones@lancaster.ac.uk*]

Developing lifelong learners: a novel online problem-based ultrasonography subject

Laura C. Minasian-Batmanian,* Anthony J. Koppi** and Elaine J. Pearson***
*School of Biomedical Sciences, Faculty of Health Sciences,
University of Sydney
**Educational Development and Technology Centre, University of South Wales
***School of Computing and Mathematics, University of Teesside

Initially published in 2000

Online learning environments have a major role in providing lifelong learning opportunities. Lifelong learning is critical for successful participation in today's competitive work environment. This paper describes an online problem-based learning approach to the creation of a student-centred learning environment for the study of the biological sciences subject in the Graduate Diploma of Applied Science (Medical Ultrasonography) course at the University of Sydney. The environment is interactive and collaborative, with all communication taking place online. Students work in groups to study clinically relevant problems. A Web-database system provides learner control in the process of knowledge acquisition, access to reference materials on the Internet and communication with the tutor and with peers through synchronous chat and asynchronous threaded discussion forums. Other online features include a protocol for problem-solving, self-assessment and feedback opportunities, detailed help, streaming audio and video and pre-course, ongoing and post-course questionnaires. This technology may be adapted to a range of disciplines and can also be utilized in on-campus teaching.

Introduction

There have been significant advances in the practice of sonography since the introduction of the University of Sydney course in 1991. The entire knowledge base can no longer be taught within the constraints of time allotted to the subject. Moreover, encouraging students to be self-directed, independent learners is more important than ever, since the field of sonography has reached a stage where there is an urgent need for new skills.

In addition, there has been a range of social and economic factors that have resulted in major changes in higher education and have contributed to the current educational climate

(Campion, 1995; Edwards, 1995). These changes have challenged universities to introduce teaching and learning strategies that cater for different client groups using forms of delivery that increase access to learning opportunities. In particular, there are demands for the university sector to provide for a larger and more diverse cross-section of the population, to cater for emerging patterns of educational involvement which facilitate lifelong learning, and to include technology-based practices in the curriculum (Renner, 1995).

These developments imply a student-centred approach to learning. Given the current economic climate in higher education and the need to maintain pedagogically sound practices, there is the need to consider learning within a developmental framework (Laurilland, 1993) which has lifelong learning as a major outcome. The introduction of problem-based learning (PBL) in the biological sciences subject of the Graduate Diploma of Health Science (Medical Sonography) at the University of Sydney in 1996 allowed students to be more independent with greater self-direction in their study programme. The subject content and structure have changed to make them relevant to the needs of sonographers in a changing professional environment. Online delivery is seen as a way of providing student-centred lifelong learning opportunities. The issue for those designing these courses is how to provide more appropriate and flexible educational opportunities for the learner, given that the importance of knowledge acquisition is the primary focus of the educational process.

This paper describes how technology was incorporated into the PBL approach in 1998 to offer enhanced learning capabilities, through the development of an innovative programme delivered entirely in distance learning mode. The online programme ran for the first time in 1999. It has attracted local, rural, interstate and New Zealand students. This accessibility has increased the size of the market, as well as making the offering of courses overseas viable. At the same time it has accommodated the time limitations of the mature-age part-time professionals of this course, without the need for simultaneous traditional university training.

Design

The multidisciplinary in-house team involved in the development of this programme over a period of six months spent 700 hours in the developmental stage of this programme at a cost of $A 22,400. The team consisted of an academic, the academic developer from the Centre for Teaching and Learning, an educational technologist and a Web developer. They had the task of collectively providing a comprehensive view of specific subject knowledge, and a pedagogical awareness of the principles underlying optimal student learning and technical skills in Web design and development. The team was conscious that the learner, not the technology, is central to the design, which, in turn, is based on sound educational principles (Godfrey, 1996). Standard Internet technology, MS SQL Server Database and Cold Fusion software were used.

The team sought to embed the fostering of generic attributes and lifelong learning skills in the program. As McAllister (1997) points out, the graduates of today are expected to be 'beginning practitioners' who become 'critical consumers of information and lifelong learners who maintain competence in their discipline, expand and test their own knowledge and skills, and contribute to the expansion of knowledge in the field'. The goals of vocationally-based education have changed from focusing on training and instruction to include an emphasis on the development of generic attributes essential to professional competence and lifelong learning (Bates, 1997; Higher Education Council, 1992; Kirkwood, 1998; Ramsden, 1992; Savery and Duffy, 1994). Engel (1995) summarized these as comprising the acquisition of:

- discipline-specific knowledge and competencies;

- generally applicable competencies, for example, ability to adapt to change and participate in change, communicate, collaborate in groups, and be self-directed lifelong learners who can apply critical reasoning and a scientific approach to decision-making in unfamiliar situations.

In order to achieve these aims, designing and redesigning content and methodology for online delivery was important, as was reconsidering previously accepted and often unquestioned practices.

The social aspect of learning present in traditional university-based education is often missing, neglected or poorly considered. Online courses often rely on simply taking written text or lecture notes and presenting them digitally. Students are expected to learn independently by reference to the written or presentation materials the lecturer would normally deliver in a classroom situation. This site, however, takes advantage of multi-media facilities to provide added value to distance learning. It combines the independence of distance learning with the social aspects of university-based learning. The site therefore contains no lecture notes, but simply presents a text description of twelve case studies that the students are required to engage with. Students work in groups of six, through the online problem identification process (Protocol) to study the pathophysiological knowledge necessary to understand and solve these cases.

The biological sciences online environment has been designed for maximum flexibility. It allows students from a variety of backgrounds to construct a study plan according to their interests, expertise and availability. An individual can opt to be delegated to research certain objectives generated by the cases. Those with particular specialties are placed together to act as resource people for the whole online community. An online calendar was incorporated to assist students in scheduling their discussions. The flexible learning approach offers the student increased choice in what to learn (students devise their own objectives); how it is learned (the learning style they choose to follow from the resources supplied); and when and where learning happens (whether it is in a face-to-face situation, at the end of a telephone line as in teleconferencing, or online conferencing).

When designing the flexible learning environment, care was taken to develop domain-specific educational strategies. These include explanatory notes in the form of text/audio hints (Figure 1) and streaming video (Figure 2), a flow chart to explain the step-by-step Protocol (Figure 3), exercises in the form of self-evaluative tests and a diagnostic exam, resources, assessment requirements and group discussion facility via chatspace and discussion forum.

The visual elements are carefully planned to take account of non-technical users. The visual and navigational interface is designed for coherence, accessibility and clarity. The

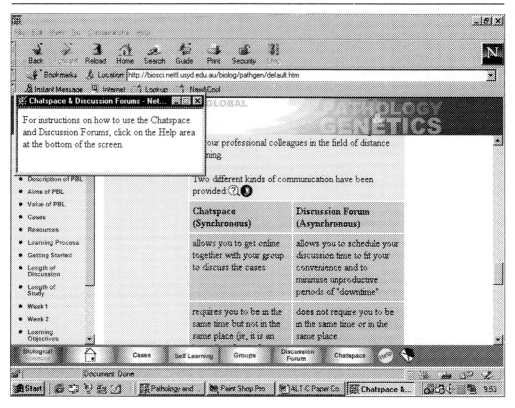

Figure 1: The pathology/genetics screen showing the use of audio hints

interface is clear with clean, crisp graphics and a subtle use of colours. The design utilizes plenty of white space, which gives a coherent feel to the display and clearly differentiates the navigational and display elements. The graphics give the impression of ultrasound without making the medical emphasis of the site too obvious. The menu structure makes the site very easy to navigate with the navigation bars remaining constant throughout. Information is arranged in three different, colour-coded sections in the site, for easy identification: pathology/genetics (blue), self learning (purple) and groups (green).

The explanation to the students of the teaching and learning methods employed is clear and comprehensive and avoids the use of jargon. Assessment is clearly explained and the course provides a spread of relevant assessment. The chat space and discussion forum provides students with the opportunity for collaborative learning online and self-tests assist students in monitoring their own progress. A series of evaluation questions provide feedback to the facilitator on most aspects of the subject and its delivery.

Self-directed learning is a feature of both traditional distance learning and university-based courses, but in distance learning it is sometimes almost the only pedagogic approach taken. Online learning can take advantage of multimedia to make a wide store of information available at the click of a button, giving all students equality of access. The use of hypertext with direct and associative links gives flexibility to the students' approach to learning. It helps to orientate the student to independence and self-responsibility in learning, a basis for

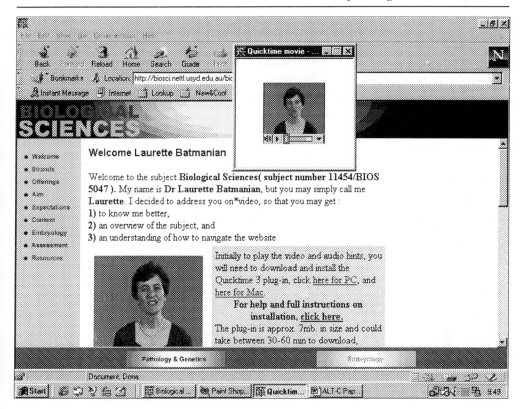

Figure 2: The biological science screen showing a streaming video

acquiring lifelong learning habits (Koppi, Lublin and Chaloupka, 1997). This site allows browsing, searching and guided discovery. The student can browse the extensive links to bibliographical references, other sites on the Internet or video clips, starting wherever they wish, following a chain of linked information intensively or skipping from one resource to another following associative links. Alternatively the student can make detailed searches using the same resources for specific information. There are also links from key words or phrases to specific and detailed information on particular aspects within the site.

The advantage of flexible delivery is that it includes learner control over content, time, place and method of learning. Online course materials are specially designed for studying when and where the students choose. The students can phone, fax, mail, or email teaching staff whenever they need help. Frequently asked questions regarding the educational aspects of the site are answered online. A suggested timetable (Figure 4), recommended length of discussions and a study schedule have been included to give participants an estimate of the time involved in this subject and the rate at which they can expect to progress. To engage students in the learning process, a case Proforma was incorporated, which summarizes the steps in the Protocol to guide the students through the learning process. Although a comprehensive online help system is provided, a fifty-page User Site Guide is mailed to students before semester starts, to familiarize them with the learning environment and navigational features.

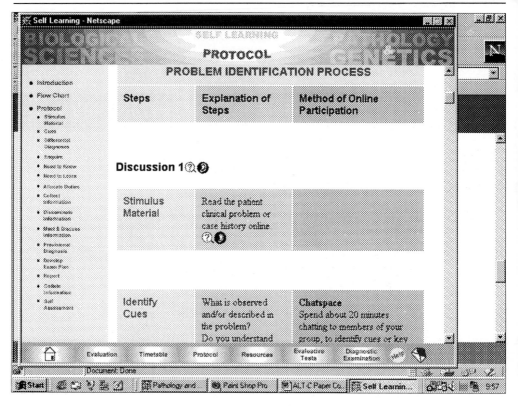

Figure 3: The Protocol flow chart

Teaching and learning methodology

Students are made aware from the outset of the teaching and learning methods adopted and are given detailed guidance on the philosophy, the features of the approach and the learning outcomes. This guidance provides early scaffolding to the students and immediately invites them to take ownership of their learning. This scaffolding is reinforced by the use of audio and video to introduce the facilitator to the student.

The entire knowledge base cannot be taught within the constraints of time allotted to the subject. Therefore, twelve cases were specifically chosen to cover the most commonly seen problems in ultrasonography practice, or those that illustrate significant principles particularly well. Clinical problems, the most common form of request presented in private practice and case histories, used to simulate the information provided in hospitals, have been incorporated. Problems were designed to begin with simple examples, becoming increasingly complex so that students can draw on previous knowledge.

Collaboration

Collaborative learning is a concept that is generally missing from traditional distance-learning courses. A particular feature of this site is the emphasis on group activity – indeed part of the assessment is based on the students' contribution to group work. Student

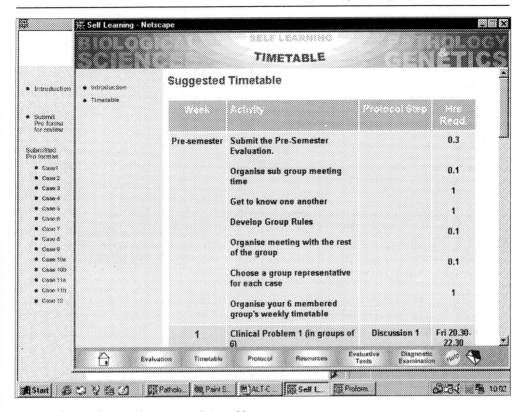

Figure 4: Screen showing the suggested timetable

success is assessed in terms of the student's ability to solve problems, communicate ideas, present information and learn how to learn, rather than simply to repeat facts. The site includes discussion groups, forums, a chat space and facilitator/student communication:

- Discussion groups are designed for small groups of students to meet online and discuss the case studies and to identify and work through the problems together. The small groups are designed to create coherent units and a group identity. This means that individuals are less likely to drop out of discussions or 'lurk' and fail to participate. Students are given advice on the structure, behaviour and timing of group discussions. Forums are designed to provide whole group discussion and interaction – an online replacement for the university-based seminar.

- There are three types of forum: (i) case studies for wider discussion of the problem, (ii) feedback where the students can discuss feedback given by the facilitator, and (iii) general discussion where students are able to raise any matters not necessarily directly related to the subject content.

- The chat space provides an opportunity for students to talk in real time (anonymously if they prefer) and to post their own opinions, references and resource links.

- Email, fax and phone communication with the facilitator are provided for support, feedback and assessment.

Feedback

In this online course, the lecturer becomes a facilitator of information (electronic, verbal, etc.) by guiding, advising, commenting on progress and offering explanations, or adopting the role of co-learner. Students are encouraged to submit alternative information with which they may be familiar from their work situation or through their literature or Web searches. This encourages the sense of interdependence and validates the knowledge that students bring to the learning situation.

The online learning environment provides the facilitator with a greater level of awareness on the progress of individuals. In the biological sciences subject we tried to engage the first-year students in a number of ways. Thirty-six students were enrolled in the subject. They were given an optional face-to-face introductory meeting with academic staff before the start of the course. Twelve students decided to learn in a face-to-face situation, twelve by online conferencing alone and twelve using a combination of the two processes. The facilitator was scheduled to attend the first four weeks of synchronous chat discussion forums, to assist students using the online environment with resolving immediate problems concerning the Protocol.

The students were initially encouraged to interact with three students of the group of six to get to know one another before semester started. They exchanged information about interests and hobbies on a personal level; their specialty at a professional level; and any other issues of common interest that they would like to share with the other students and the facilitator. Online participant profiles, which are accessible and amenable to change, were provided to personalize the design of the site (Pillay et al, 1998). This extracurricular activity was used as a focus to acquaint the students with their peers and to introduce them to the new learning environment. This was found to be beneficial before students worked in groups on their clinical problems or case studies. The groups were provided with guidance to help them devise criteria for working effectively as a team. This reflects the skills they will require in their future professional and social lives. The proportions of both peer and facilitator support have been increased to four hours per week from the normal rate of two hours per week. This is intended to overcome feelings of isolation and to allow for the fact that learning and motivation are significantly improved for distance-learning participants by increased communication and feedback. The facilitator was always on duty to respond to email, fax or phone, although in the future this will be restricted to certain hours in the week.

Assessment

The online subject features both formative and summative assessment. Formative assessment is provided through a series of self-evaluative tests at various stages in the course and a diagnostic mock examination. Self-tests enable the students to monitor their own progress in a non-threatening situation and thereby take control of their own learning. The use of formative assessment through questionnaires or short-answer questions means students are able to test their own ideas against 'expert' opinions. These tests are also useful in assisting students in the identification of learning goals and areas of difficulty and provide a means of reflecting on their on-going progress. Other formative assessment comes from peer group discussion and feedback from the facilitator.

Summative assessment emphasizes the importance of collaborative learning, with a proportion of the marks based on contribution to group work (5 per cent), as well as ten reports submitted by the group (40 per cent). Individual assessment (28 per cent) is based on an end examination comprising multiple-choice and short essay questions. Students are provided with additional scaffolding in the first clinical problem, which is presented as a worked example and hints are provided to help solve case history two, giving an example of the scope, focus and level of work required for the assessed cases.

Evaluation

Staff at the University of Teesside, UK, evaluated the biological sciences subject. The evaluation was based on an examination of the features that contribute to the PBL approach as well as visual, content and navigational design elements. After evaluation, the subject was presented to the students as an example of best practice in online learning. The results of the evaluation have been incorporated into this paper.

This online environment enables integration and co-ordination of theoretical and clinical material, as well as encouraging the development of skills in logical reasoning, critical thinking, communication and self-directed learning (Grabinger *et al*, 1997). Biggs (1989) found that teaching that gave evidence of deep learning contained one or more of the following: an appropriate motivational context; a high degree of learner activity; interaction with others, both peers and teachers; and a well structured knowledge base. The students were highly motivated through the direct link between the cases and the demands of the subject and its assessment. There was a high degree of learner activity with brainstorming, role-play and feedback (Moore, 1993). Students had to interact constantly with their peers and facilitators. We believe that the programme described in this paper has led to deep learning, as evidenced by their results of 94 per cent graded passes this year, as compared with 84 per cent last year. Although evaluation of student responses is yet to be fully analysed, preliminary feedback has been most favourable. Some representative comments included:

'The best aspect of the subject was the problem-based learning approach.'
'The tie-together of pathology and its effect on the patient and what we should expect to observe with ultrasound made the subject relevant.'
'Diagnosing the case studies was interesting. It was good to be in groups as we got to know other people in more depth.'
'Links were very helpful even now after bioscience has finished.'
'My skills in written communication as well as the use of technology was greatly improved.'

Discussion

Consistent with the philosophy of critical and reflective practice, we have attempted to accommodate the attributes of lifelong learning and critical reasoning skills development, described by Candy (1994), to the study of ultrasonography. The cases require that the students:

- ask questions of the facilitator and of each other to determine objectives, undertake critical appraisal of the relevant literature, and self-evaluation of understanding;

- are aware of the interrelationships between theoretical and applied knowledge and acquire an understanding of ultrasound procedural limitations;

- develop a knowledge of current resources, frame researchable questions, locate, evaluate, manage and use information in a range of contexts, retrieve information using a variety of media, including Web links, videos, literature searches; decode information in a variety of written and graphical forms; and critically evaluate information to be submitted in the report;

- are capable of working on their own, interact within a team, and develop self organizational skills as well as an appreciation for multiple perspectives;

- acquire a variety of learning skills, an awareness of their own personal strengths, weaknesses and preferred learning style; develop different strategies for learning according to context and understand the differences between surface and deep learning.

Conclusion

This paper has shown that technology can assist in the adoption of a more student-centred approach to teaching and learning in which students develop knowledge, skills and resources that can assist them to a stage of 'learning by discovery', through a process of discussion, discovery, interaction, adaptation and reflection. It has the advantages of providing increased accessibility, immediate feedback, interactive learning and a more flexible environment. It provides both peer and tutor support and guidance, as well as a wealth of resources. The subject encourages active learning, without leaving the student awash in a sea of information, by providing guidance on how to begin addressing the problems introduced in the case studies. It provides a complete environment from introduction to assessment and evaluation.

Another advantage of this site for the courseware developer is that the format adopted is easily transferable to a different subject and course, making reuse a practical proposition and thereby cutting down on future development. The template is currently being adapted to accommodate for the design and development of a postgraduate subject in Management in the School of Health Information Management, Faculty of Health Sciences, University of Sydney. Moreover, active interest has been expressed for use in undergraduate courses run by the Department of Biological Sciences and continuing education courses run by the Department of Obstetrics and Gynaecology, at the University. The site has been adopted by the University of Teesside as an example of good practice in teaching courseware design at Masters degree level.

Acknowledgements

Laura Batmanian wishes to thank Steve Clark and Andrew Lovell-Simons for their assistance in designing and developing the site; Jason Bayly for Web design, programming and co-ordination; and Angus Denton for the productions of the streaming video and audio hints.

References

Bates, A. (1997), 'The impact of technological change on open and distance learning', *Distance Education*, 18 (1), 93–109.

Biggs, J. (1989), 'Approaches to the enhancement of student learning', *Higher Education Research and Development*, 8 (1), 7–25.

Campion, M. (1995), 'The supposed demise of bureaucracy: implications for distance education and open learning – more on the post-Fordism debate', *Distance Education*, 16 (2), 192–216.

Candy, P. (1994), 'Developing lifelong learners through undergraduate education', Commissioned Report No. 28, National Board of Employment, Education and Training, Canberra: Australian Government Printing Service.

Edwards, R. (1995), 'Different discourses, discourses of difference: globalisation, distance education and open learning', *Distance Education*, 16 (2), 241–55.

Engel, C. E. (1995), 'Medical education in the twenty-first century: the need for a capability approach', *Capability*, 1 (4), 23–30.

Godfrey, R. (1996), 'The World Wide Web: a replacement, displacement, supplement or adjunct of traditional methods', in James, C. P. and Vaughan, B. (eds.), *ASCILITE 96: Making New Connections, Proceedings of the 13th Annual Conference of the Australian Society of Computers in Learning in Tertiary Education*, University of South Australia, Adelaide, 221–34.

Grabinger, S., Dunlap, J. and Duffield, J. (1997), 'Rich environments for active learning in action: problem-based learning', *ALT-J*, 5 (2), 5–18.

Higher Education Council (1992), 'Higher education: achieving quality', AGPS, Canberra, 20–2.

Kirkwood, A. (1998), 'New media mania: can information and communication technologies enhance the quality of open and distance learning?', *Distance Education*, 19 (2), 228–41.

Koppi, A. J., Lublin, J. R. and Chaloupka, M. J. (1997), 'Effective teaching and learning in a high-tech environment', *IETI*, 34, 245–51.

Laurillard, D. (1993), *Rethinking University Teaching: A Framework for the Effective Use of Educational Technology*, London: Routledge.

McAllister, L. (1997), 'Towards a philosophy for clinical education', in McAllister, L., Lincoln, M., McLeod, S. and Maloney, D. (eds.), *Facilitating Learning in Clinical Settings*, Cheltenham: Stanley Thornes, 214–45.

Moore, M. (1993), 'Three types of interaction', in Harry, K., John, M. and Keegan, D. (eds.), *Distance Education: New Perspectives*, London: Routledge, 19–24.

Pillay, H., Boles, W. and Raj, L. (1998), 'Personalising the design of computer-based instruction to enhance learning', *ALT-J*, 6, 17–32.

Ramsden, P. (1992), *Learning to Teach in Higher Education*, London: Routledge.

Renner, W. (1995), 'Post-Fordist visions and technological solutions: educational technology and the labour process', *Distance Education*, 16 (2), 284–301.

Savery, J. R. and Duffy, T. M. (1994), 'Problem based learning: an instructional model and its constructivist framework', *Educational Technology*, 8, 31–8.

Contact author

Dr Laura Minasian-Batmanian is a Senior Lecturer in the School of Biomedical Sciences, Faculty of Health Sciences at the University of Sydney. Her current research interest is in researching, developing and implementing problem-based learning and technology for the benefit and enhancement of student learning in the faculty.
[*l.batmanian@cchs.usyd.edu.au*]

The future

Real-time interactive social environments: a review of BT's Generic Learning Platform

Michael Gardner and Holly Ward
Internet and Multimedia Application, BT Adastral Park,

Initially published in 1999

Online learning in particular and lifelong learning in general require a learning platform that makes sense both pedagogically and commercially. This paper sets out to describe what we mean by generic, learning and platform. The technical requirements are described, and various trials that test the technical, educational and commercial nature of the platform are described. Finally, the future developments planned for the Real-time Interactive Social Environments (RISE) are discussed.

Introduction

The vision of a single platform that supports all the key stakeholders within a wide range of distance learning activities has long been a dream of educationalists and commercial organizations alike. Such a platform could not only provide a consistent educational 'experience' for teachers and learners throughout their lives, but it could also provide administration and managerial support, all of which can be tailored to individual requirements.

This paper reviews the work carried out within BT's Education and Training Research Programme over the past three years in developing a generic platform for online distance learning. The first section of the paper reviews the original objectives of the Generic Learning Platform (GLP) and the context of the education and training research within BT. Following this, a description of the Real time Interactive Social Environment (RISE) platform is provided including the technical components of RISE.

Over the past three years the research team have carried out a number of trials using RISE. Each of these intended to stretch the overall architecture and our own understanding of what enables successful distance learning across a range of educational settings. An overview of each of these trials is given and the implications for implementing a GLP. Finally this paper closes with some comments on the future of RISE.

The meaning of generic

In the context of our work, we define 'generic' as a solution that is not dependent upon any particular individual's learning stage, be it at school, college, university or work. Such a definition is essential to support meaningfully the notion of lifelong learning. The concept of learning being generic has two important implications. First it assumes that there is greater commonality rather than differences at the developmental stages of learning throughout most of our lives. The learning process is essentially universal, although it may require degrees of customization to meet some of the specific needs of the learners at their different developmental stages, situations or locations. Many of these learning issues will be addressed in the next section.

Generic solutions also make considerable commercial sense. Solutions specifically designed for each sector (i.e. schools, colleges, universities and learning for leisure) are expensive and may not facilitate movement between sectors as required by a lifelong learning scenario. A generic solution recognizes a common core which may be customized to suit the needs of a sector (see section on the platform for more details on customization). Finally the term 'generic' is also making a statement about content that has both educational and commercial implications. Content is the critical component that needs to change to reflect the differing educational requirement – it cannot be generic across sector, at best it is common for a single homogeneous group of learners. The GLP, with respect to content at least, must be considered as a generic 'shell' into which appropriate content can be added. Commercially this also makes sense as BT does not consider itself as a content provider and works with third parties who have a high reputation in the area of content provision.

Our aim therefore was to produce a platform that was sufficiently generic to allow the 'reuse' of basic or core components across the market sectors, thus keeping development costs down, and sufficiently universal in terms of learning to provide a shell that requires the minimum of customization. To achieve this aim requires a conceptual framework that serves as a bridge between the theoretical academic educational research and the practicalities of implementing real distance learning systems. It can also provide a consistent framework for the design of systems and an evaluation framework in which to assess them. A true test of the generic nature of the platform is best assessed through empirical trials. Trials of the GLP in the higher education, further education and school sectors are described in later sections.

The meaning of learning

The theoretical background and underpinning for the GLP was provided by Mayes's 'learning framework' (1994). The original Mayes framework (1994) is illustrated in Figure 1 and can be summarized in the following way:

- Conceptualization: this refers to the student's initial contact with learning material. This could be achieved by attending a lecture or seminar, reading textbooks, watching television, listening to the radio or by using multimedia material on the Web.

- Construction: at this stage students build on the concepts learnt in the conceptualization phase and refine their understanding by working on further tests and examples. This could be achieved by carrying out laboratory experiments or working on practical homework, or using multimedia learning materials.

- Dialogue: at this stage students refine their understanding through dialogue and discussion. This could be achieved by participating in tutorials or having informal and sometimes impromptu conversations. On the Web this could be supported by using shared whiteboards, conferencing tools (e.g., audio, video, data conferencing etc.) more complex shared spaces, or discussion groups.

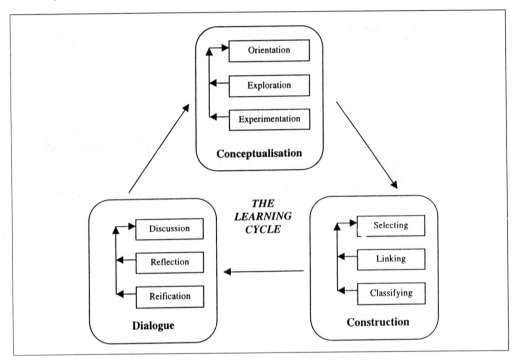

Figure 1: Mayes's conceptual framework

In general within the sphere of commercial distance learning platforms there has been much greater emphasis on supporting conceptualization, much of it through the development of multimedia Web and CD-ROM content. There tend to be fewer tools to support construction and dialogue as part of an overall educational process. It is for this reason and also the suitability of the constructionist approach to distance learning, i.e. a learner-centred approach, that much of the team's research is aimed at supporting construction and dialogue within this overall educational framework. The key is to identify and support the complete end-to-end 'education process' rather than focus primarily on the end-products of education, e.g. teaching and learning content. The initial requirement therefore for the GLP was that it would support the three-stage Mayes model.

From a commercial perspective the GLP is attractive to BT because it could provide a single service platform which could be used to provide tailored services to different market sectors, i.e. higher and further education (HE/FE), schools, home etc. In this context the platform would need to support a much wider range of functions than just those covered in Mayes's model. A service definition for the FE sector, as shown in Figure 2, illustrates this.

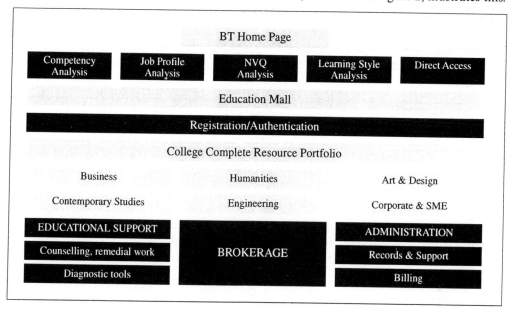

Figure 2: A typical service definition for the FE sector

In this diagram the service definition is made up of five key components:

- Access point: in its simplest form this could be a homepage, but could be more complex if considered as part of a larger range of online educational services.

- Analysis and profiling: tools need to be provided to assist a student or teacher to gain access to the appropriate online resources. For example this can include competency, job profile, or learning style analysis, or it could be based on a defined curriculum, e.g. a National Vocational Qualification (NVQ), or direct access to online resources using advanced search and retrieval tools.

- Administration: this can be the most important component for educational institutions, and will include the user registration and authentication process, tracking, support, billing and content provision.

- Online resources: there will need to be a method for classifying online learning resources and structuring the methods of access. Also in its widest context, online resources can include multimedia content, live and pre-recorded lectures, real people, e.g. tutors, and also resources generated by the students themselves.

- Support functions: these will include a range of support functions such as access to tutors and mentors, counselling and remedial tools.

The BT generic learning platform therefore needed to support not only the Mayes conceptual framework, but also other functions, as defined above, that are necessary to provide a complete online service.

Development of the RISE platform

RISE is the instantiation of the GLP that we have developed at BT. Its development has been an iterative process based on a series of trials. These trials were deliberately chosen to stretch the platform and so lead to a better understanding of the issues involved in developing a GLP. This section describes the development of the RISE GLP through research trials.

RISE architecture

The RISE system had to provide the following core functions:

- Personalized access to online information based on the online group of students, tutors and support staff. Each person would have a personalized information space representing their role within the group and the actual work that they had each completed.

- A virtual classroom or meeting-place which could provide access to good-quality multi-party audio conferencing between participants, and the ability for users to share information and record their conferences.

- Public and private information spaces where students could store their audio-recordings and coursework.

- An online shared diary to enable participants to plan their schedules online.

- Access to a modular set of coursework based on defined stages, modules and checkpoints, plus tools to enable the teaching staff to update and manage the course materials.

- Email and text discussion groups.

- Personalized home pages.

To meet these requirements RISE required three essential features: the integration of an audio conferencing capability with the Web; a dynamic database-driven Web server which could be personalized to each individual user; an online dynamically updating Web Meeting Place. The RISE architecture is illustrated in Figure 3.

RISE is designed around the combination of a data-driven Web server and an audio-conferencing server. The Web server is a standard Microsoft Internet Information Server (IIS) on a Windows NT platform. All of the RISE application software is written in Apple WebObjects. WebObjects builds the Web pages dynamically according to the business rules in the RISE application and the underlying data in the Oracle database. In addition to the RISE application there is a separate ConfMan application that manages all conferences and acts as an interface to the Aculab Millennium audio-conferencing platform. The Aculab Millennium is connected to two ISDN–30 lines that will support at any one time up to 60 callers in conferences. Also there is a separate Java server application called REEL. This provides a Web page-sharing capability to users in a conference. Using a simple Java Applet conference participants can push Web pages to each other's browsers. Users of RISE only require a Java-enabled Web browser, e.g. Internet Explorer 3+ or Netscape 3+. No additional client-side software is required.

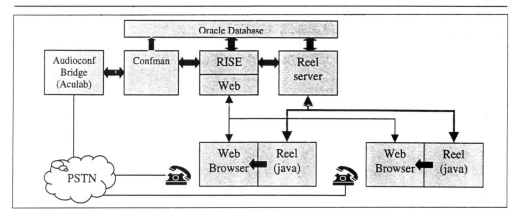

Figure 3: RISE architecture.

The key to RISE is the underlying database and the structure of the data schema. Figure 4 shows the data schema used in the Merlin trial. In many ways the success of the GLP will be dependent on the richness and flexibility of this data schema to support a range of different distance learning scenarios. Irrespective of the other technologies used, the data schema defines the distance learning model for the GLP and should be at the core of any new online service.

RISE is built around the ability to create any number of separate closed user groups. Using a number of Web-based online forms a system administrator can create and configure a closed user. For each group, the administrator must configure the group, the roles of users in the group, e.g. tutor, student, administrator, etc. and the layout and functions available to the group. In this way RISE can be used to create a customized system to support a closed user group of individuals. By using the database to build dynamically the online pages for each user it is relatively easy to administer the RISE groups based on the database schema and the customizable options available in RISE. Merlin represents one such closed user group and the Merlin requirements influenced the development of the first version of RISE. However, it was recognized from the start that any GLP must support a large range of different course and learning models, which is why a range of different trials were carried out in order to extend the capabilities of the RISE GLP.

RISE interface and functionality

The user must first log into RISE with their username and password and the group that they belong to. Once successfully logged in, the user will be presented with two windows, see Figure 5. The larger window contains a toolbar and a workspace area. The toolbar buttons and the 'look and feel' of this window are all dynamically configured for each group. The smaller window is the Web Meeting Place and allows users to see who else is online in their group, enabling them to initiate and take part in audio-conferences with other users. Once in a conference, users can record their conversation, invite other people, and control the privacy of the conference. They can also share Web pages with one another using the REEL Applet, see Figure 6. BT has secured two patents on the RISE system concerning the way the Meeting Place is used to control audio conferences and the use of the REEL application to share Web pages. RISE also provides email and discussion group facilities for all users within the group.

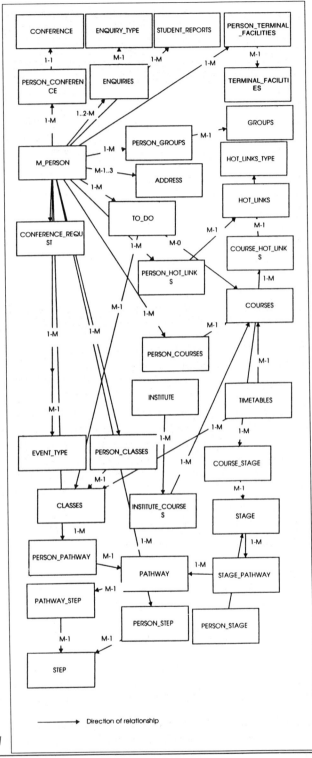

Figure 4: The Merlin Data Model

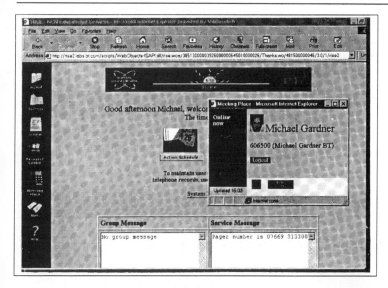

Figure 5: RISE screens

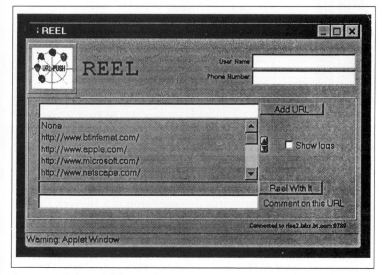

Figure 6: Reel Applet
Window

As part of the RISE core functionality each user has access to their own private folder and a shared folder which is available to everyone in the same group. These folders can be used to store any work completed by individuals in the group. Work items can include audio-recordings from RISE conferences or any uploaded computer file. Users can easily upload files from their own PC into these folders for their own use, to share with the tutor, or to share with the whole group. When audio-conferences are recorded the system converts them automatically into RealAudio format. This allows users to stream the audio file back to their PC when they want to listen to it, rather than downloading a large file to their PC. Users can also email the URL of the recording to another person.

In addition to the RISE Meeting Place, the students and teachers also needed a way of scheduling when they would be online, and organizing conferences with one another. It is possible to organize conferences formally using asynchronous tools such as email. However, this can be inappropriate if students or teachers want to organize informally their 'online time' to ensure that they will be online when others in their group are also online. To meet this need we developed a shared online diary in RISE, see Figure 7. This allowed users to publicize when they intended to be online. Students could organize their use of the system when others would also be online. This type of informal groupware support is crucial to the success of any distance learning group.

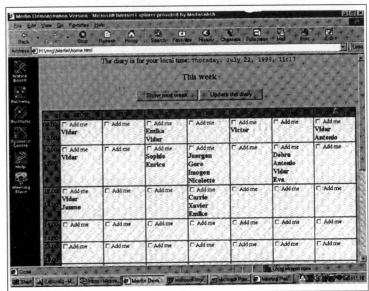

Figure 7: The Merlin Diary

The testing of RISE

The first use of RISE was in the Merlin trial (Miller, 1996) which ran from October 1996 through to June 1997. The Merlin project was a collaborative project between BT and the University of Hull Language Institute. It was primarily designed around a 16-week English as a Foreign Language (EFL) course delivered online to 40 students world-wide. Merlin was influential as it defined the requirements for the first version of RISE and provided the core features for all subsequent trials. Merlin also set the model for future collaborations, whereby BT acted as a service provider of RISE to an educational organization. Through these trials we learnt a lot about how this relationship could work effectively, how to facilitate the understanding of requirements and how to enable essentially non-technical teaching organizations effectively to deliver courses online via the Web.

Merlin required the following core features:

- Conceptualization: a 16-week modular course with assessment points. The content for the course would be released at set periods under the control of the tutors.

- Construction: students had to be able to practise their English with a peer group and tutors, be able to record their spoken work and share it with others on the course. They had to complete a range of online exercises and save results as well as access standard course materials.

- Dialogue: students and tutors had to be able to hold both formal and informal meetings using audio conferencing and be able to share information online.

The Merlin trial had distinct requirements in terms of the provision of online courseware to the students. The EFL course was based on a 'pathway' of eight separate stages with checkpoints at the end of each stage, e.g. a formally assessed exercise. Within each stage were a number of separate modules that students could work through to enable them to complete the stage. The key for RISE in this scenario was to enable the course deliverers, Hull University, to manage their courseware and the extrinsic course process. A number of course management tools were developed to meet this end. These tools allowed the Hull team to add dynamically new modules and exercises using bespoke administrator forms. In this way the course material could be released over the 16-week period and the tutors could dynamically update the material based on students' progress and lessons learnt from the students' use of the previous courseware. We also provided the Hull team with a basic Applications Programming Interface (API) which allowed them to develop course material which included online exercises. The API allowed the courseware team to define these exercises in a way that allowed the students to store their answers to the exercises in the Merlin workbook areas, and index these answers against the specific part of the EFL pathway that they came from. In summary, these tools allowed the course providers to have complete control over the online courseware and be able to build assessable course exercises, which could be tracked against each student. In addition to using these tools, any Web-based content could also be linked into the system as courseware or as an online resource.

The Merlin trial had fairly specific requirements for the EFL course based on a strict modular course structure and a rigid curriculum. This type of formal course structure is fairly commonplace. There are many similarities to other school and HE-based curricula. However, particularly in vocational courses there is a need to support a much more flexible course structure, one where the emphasis is more on the learner to assemble evidence of particular competencies that can then be assessed by a tutor, e.g. a student undertaking an NVQ. In 1997, BT ran a second trial using the RISE platform in collaboration with Northern Colleges Network (NCN), a group of FE colleges in the Northumberland region. In addition to testing the core technologies, the key objectives of this trial were to evaluate the usability of a GLP in the FE sector, increase awareness of online learning throughout the FE sector and provide early indications of the acceptability of technology for online learning. In addition to this the team investigated the effects of organizational factors on implementation of online learning (Ward *et al*, 1998).

During this trial the BT team worked closely with the NCN tutors and administrators to develop a version of RISE to deliver the NVQ Level II in Information Technology. The key requirement for the RISE platform in this instance was the need to support the NVQ process. To meet this end, a number of enhancements were required. This included an NVQ Schedule Management Tool to allow students to monitor and update their progress against all elements of their NVQ. Tutors could also use the tool to oversee and monitor the students' progress and update the schedule when necessary, e.g. following formal assessment.

Another requirement from the NVQ process was fully to support the collection and assessment of student evidence documents. In Merlin, Personal and Group 'workbooks' were provided to allow users to store items such as recordings of conferences and completed coursework. The NVQ process required that the system could more formally support the process of capturing, sharing and assessing evidence material. To meet this need four 'workbook' areas were created:

- an 'audio-evidence' folder, where students could save private recordings from audio conferences;

- a 'meeting records' area, where users could save shared meeting recordings;

- a 'progress review' area, where students could upload documents related to the formal NVQ progress reviews;

- a 'text evidence' folder, which could be used by students to maintain a store of all their documents related to the evidence required.

The access privileges to all of these folders were managed by the RISE system to ensure that students only had access to their own records, and tutors could access all records for all of their students.

So far the RISE platform had only been used to support formal distance learning courses based on a curriculum of task activities and formalized assessment points. The next key question was whether the GLP could be applied to less formalized learning communities, where rather than supporting a set curriculum the GLP needs to support the general communications and learning needs of the community. To investigate this scenario the HomeLearn trial was initiated. This was a collaboration between BT and Apple Computer Inc. and used Wickham Market Primary School in Suffolk as the case study (James, 1999; Tracey *et al*, 1999). The aim was to investigate how RISE could be used to support the relationship between home and school. Students from Years 5 and 6 and their families participated in the trial. They were provided with a PC and an Internet connection at home. RISE accounts were created for teachers, students, their families and a school governor. No additional 'content' was created to support this trial as an explicit course was not being run. Instead the team investigated the communication needs of participants. To meet this requirement RISE was extended to include facilities for:

- Tracking and managing students' homework. The teacher could create, assign and monitor homework for the trial students. Students could track their homework and parents could monitor their child's homework and electronically authorize it as being 'seen'.

- Publication and sharing of school's long- and medium-term plans. A tool was developed to allow the teacher to create daily short-term plans and publish them for the trial community.

- Discussion group areas were extended to allow a greater variety of discussion areas. Some were only open to distinct user sub-groups, e.g. children-only or parent-only. This also included project-specific individual and group folder areas, which could be used to support specific activities.

- The RISE conferencing facility was extended to allow invited experts to attend specially organized conferences as part of the students' project work, e.g. a meteorologist in the geography class.

A further RISE trial was conducted in collaboration with the University of the Highlands and Islands (UHI) in 1999. This used the RISE general purpose group and conferencing facilities to support the delivery of online lectures to a group of students undertaking a software engineering B.Sc. course. Students and lecturer would log into RISE and join together in an audio conference using the RISE Meeting Place. Once in a conference, the lecturer would use the REEL tool to 'push' lecture slides to each student's Web browser whilst giving the lecture as an audio conference. In addition to the UHI trial with RISE, the BT team have developed a number of other tools as part of an Advanced Learning Environments (ALE) project to assist in the capture and synchronization of slides and audio and in archiving and distributing lecture material (see Mudhar and Fowler, 1999).

Our final trial is ongoing and involves working with Suffolk College on a European Social Fund (Adapt) project, COVETS, to develop a 'Televersity' platform for Suffolk. At the heart of the system will be *Solstra* (*http://www.solstra.com/*), which will provide all of the user registration and content provision aspects. RISE will be integrated into this system to provide support for the community conferencing and the upload and sharing of individual and group work. The plan is to test this platform through some early trials based on specific courses, but widen this out to address many of the issues raised by the UK Government University for Industry (UFI) initiative. A particular focus will be on providing education and training to meet the needs of a wide range of different people in the community and laying some of the foundations for the support of lifelong learning.

This section has described the development of the RISE Generic Learning Platform based on a number of trials with different educational contexts. The next section highlights some of the key findings from the evaluation of some of these trials.

Key findings from trials

System usability

As with any experimental software the trials encountered some technical problems. Those highlighted in the early trials were overcome and RISE was refined for the following trial. The key problem, identified during the Merlin trial, surrounded the issue of integration. Users were uncomfortable using several applications at the same time and frequently requested a combined user interface from which to launch all applications. To resolve this *FirstClass* and RISE were integrated into the same browser for the NCN trial. The result was positive feedback from the users. This demonstrated that RISE was a tool with an acceptable level of hidden technology, the user simply logged on and completed a task without identifying that they were using two different applications.

Many of the users expressed a desire to have multi-party video conferencing facilities built in to the RISE platform in addition to the audio conferencing tools already available. Prior research had indicated that users were less interested in 'seeing' one another than they were in actually completing the task at hand, e.g. sharing a Web page, and preferred better quality audio and faster data-sharing facilities than a video image. During the UHI trial BT carried out a comparison of lectures delivered using video conferencing to those delivered over the audio link built in to RISE. The students reported that the quality of the video conferencing was very poor but they still felt that actually being able to see their lecturer was beneficial. However, the lecturer suggested that video conferencing 'didn't add a huge amount over audio conferencing'. He was comfortable using RISE to deliver and support his slide material. Despite students' early concerns, at the end of the trial all of the students 'preferred using RISE to the video conferencing' (Tracey, 1999).

It should be noted that students who were already friends or had met their tutor prior to the trials were comfortable conferencing without a video image. Only those who had no previous contact were initially uncomfortable with an audio link alone. Many users commented that making contact with a complete stranger is much harder in a 'virtual' environment than it is in a 'real' classroom. However, as the users familiarized themselves with the system, the course and their peer group, their confidence grew and their need for the visual cues changed. This highlighted the need for a coherent induction process to any course delivered online.

All users used the email actively and found this to be of huge benefit. Students on the UHI trial found this useful as the lecturer was not always online for synchronous communication, but by leaving a text message using *FirstClass* the students were confident that they would receive a prompt response: 'It was very convenient getting in touch with him [the lecturer] by email.'

On a similar BT research collaboration based in Essex, the School Centred Initial Teacher Training (SCITT) course, tutors actively encouraged remote PGCE students to submit assignments using *FirstClass*. This was eventually made compulsory and the course manager identified several reasons for this, including the ability to save and file the assignments on the *FirstClass* server, establish group folders with restricted access for security, and the ability to trace the history of messages. Participants on the NCN trial identified the same benefits.

Teaching staff experiences
By implementing an online service to deliver and manage a course or series of courses the teaching staff found that their roles changed slightly. They highlighted the issue that they were now expected to be first-line technical support for the student groups regardless of their main subject area. Educational institutions need to review this requirement prior to the implementation of a RISE-type platform to deliver or enhance any course.

However, technical support aside, estimates from the workload of staff, specifically from the Merlin trial, indicated that a tutor could support a larger student group online than in a classroom-based course with a similar input of hours. However, educational institutions must be aware that an increasing number of students increases the level of administrative support required.

Staff also found that they built stronger working relationships with students during the trial. In particular during the Merlin trial the tutor's role was that of a facilitator rather than course leader. Due to this and the increased opportunities for one-to-one online contact the tutor commented that 'there were more opportunities to get to know students'.

The feeling of being better able to address the individual needs of students was common amongst staff in the NCN, Merlin and HomeLearn trials.

Students' feedback
For all students who took part in the RISE trials the concept of delivering or participating in an online lecture was a new experience. Many of the staff and students expressed

concerns about what to expect from the system. Even students who considered themselves to be technically competent were wary of how the trial would run. For example one of the students from the UHI trial commented: 'It was a bit of the unknown really. To be quite honest I was a bit unsure about it, I just didn't think it would be as good as somebody standing there in front of a blackboard.'

However, the advantages of a GLP, in particular the flexible working opportunities, soon outweighed the initial concerns. All the students who participated in the UHI trial wanted to use a RISE-type platform again. While this was an experimental platform, the students were undertaking a real course and they achieved 100 per cent pass rate while using RISE.

One of the early questions in all these trials addressed the issue of student-tutor interaction. How would it be affected by the technology? The majority of HE/FE students considered the amount of support or contact with their tutor or lecturer compared favourably with that from classroom teachers. For Merlin students the spoken contact with the tutor was more enjoyable and beneficial than in classrooms. For the students participating in the NCN trial the ability to send and receive emails and book online meetings with their tutor was essential and they felt they received better attention than if they were in a group tutorial or meeting. NVQs usually involve less contact with a tutor and according to the students the introduction of RISE had improved this level of interaction.

Learning gains

The objectives for the students during the NCN trial included using RISE to download courseware related to their NVQ, begin the exercises and start to gather evidence towards their NVQ. Three-quarters of the students had downloaded courseware and one-third had reached the stage of submitting evidence. Over 80 per cent of the student group felt that they had benefit in terms of ICT skills, knowledge of GLPs and general confidence levels using a GLP (Ward *et al*, 1998). Of the RISE trials this was the only instance where students participated voluntarily and the trial was in addition to their work and study commitments. All other trials were delivering 'live' courses.

The use of RISE in the community environment during the HomeLearn trial increased the students' access to resources. For example, an audio conference with the author of their current English textbook was arranged after school as a homework assignment. The children had the opportunity to ask the author specific questions and engage in spontaneous debate following the answers. This and other similar activities not only enabled the students to complete their homework but gave them the opportunity to develop ICT skills, presentation, interpersonal and communication skills and research skills. They could achieve all this from their own homes with the support of either friends or family as well as receive the support and guidance from teaching staff in the formal classroom environment.

Conclusions

A GLP works very well as the sole learning environment in a situation whereby the student is self-motivated and does not consider the teaching staff to be the sole or principal learning resource, e.g. Merlin or NCN trials. In this case students manage the majority of their learning experience. As a communication tool used to bring together communities that ordinarily would have less contact, e.g. the school and parents in the HomeLearn trial, RISE is very successful but would not replace the formal classroom-based environment that young children require. It can provide an excellent gateway to exploring a global resource bank, via the Web and email, to supplement and support the National Curriculum, but should not be considered as a replacement for the teacher during compulsory years.

The future

This paper has reviewed the work done in BT to develop a GLP through the development of RISE. BT is now developing a commercial platform that delivers many of the features that were previously provided by RISE, and work is continuing, e.g. COVETS, to extend features and coverage of the new commercial platform.

With the growing maturity of this area, international standards will have a much larger part to play in ensuring that there is consistency in all aspects of distance learning. In particular, the Instructional Management Systems (IMS) Project (1999), an Educom NLII initiative, is developing a specification and software for managing online learning resources, where learning resources can include people, educational service companies, content, tools and activities. With the rapid increase in the availability of online educational content a common methodology and language for describing online learning resources is required. Content can often be described in many different ways. These descriptive categories are meta-data about the object. Online materials need a similar system of meta-data.

Standards are also emerging for the design of the generic learning platform itself, notably the IEEE Learning Technology Systems Architecture (LTSA) as described by Farance and Tonkel (1998). This Architecture Specification describes the high-level system design and the components of the LTSA. The LTSA specification covers a wide range of systems, including learning technology, computer-based training, electronic performance support systems, computer-assisted instruction, intelligent tutoring, education and training technology, and meta-data. The LTSA specification attempts to provide a framework for understanding existing and future systems, and promotes interoperability and portability by identifying critical system interfaces such that any system can comply with the LTSA.

BT is continuing to carry out research into GLPs, in particular:

- investigating new types of content to support collaboration and construction;

- addressing psychological issues of distance learning, e.g. how to engage learners, and new literacies required by learners;

- understanding the needs of complex learning communities and mapping these through to the design of online support systems.

In BT, RISE has provided us with a rich understanding of how distance learning can be effectively supported and through a variety of trials has shown how a single GLP can support a range of different learning communities.

References

Fowler, C. J. H. and Mayes, J. T. (1997), 'Applying telepresence to education', *BT Technology Journal*, 14, 4, 98–110.

Farance, F. and Tonkel, J. (May, 1998), 'Learning Technology Systems Architecture (LTSA) Specification', Version 4.00, A Base Document for IEEE 1484.1, *http://www.edutool.com/ltsa*.

James, G. (April 1999), 'Inviting the world into your school', *Interactive Journal*, 27–9.

IMS Meta-data project (October, 1999), *IMS Meta-data Specification*, *http://www.imsproject.org/metadata/*.

Mason, R. (April 1998), 'Models of OnLine courses', *Networked Lifelong Learning (NLL), '98 Conference Proceedings*, 1.72–1.79.

Mayes, J. T. (1994), 'Learning through telematics', *A Report Commissioned by BT Report*, © BT Telecommunications Plc.

Miller, C. (1996), 'Initial user evaluation of Merlin', *BT Internal Report*, © BT Telecommunications Plc.

Mudhar, P. and Fowler, C. J. H. (1999), 'The advanced learning environment (ALE) demonstrator: from theory to design', *Computer Aided Learning (CAL) '99 Abstract Book*, 49.

Smythe, P. and Gardner, M. (1997), 'The RISE Platform: supporting social interaction for on-line education', CHI 97 Electronic Publications: *Late-Breaking/Short Demonstrations*.

Tracey, K. (June 1999), 'UHI evaluation report', *BT Internal Report*, © BT Telecommunications Plc.

Tracey, K., Fowler, C. J. H. and Penn, C. (January 1997), 'Developing an infrastructure to support communities of learning', *BT Technology Journal*, 17, 1, 98–110.

Ward, H., Tracey, K. and Barker, M. (April 1998), 'Developing a virtual college', *Networked Lifelong Learning, '98 Conference Proceedings*, 4.43–4.51.

Update – Real-time interactive social environments: a review of BT's Generic Learning Platform

Since the completion of the paper there have been a number of notable developments. The first is the development work on the COVETS project. This has involved the integration of some of the core components from RISE with BT's commercial training platform *Solstra*. The focus of the COVETS Televersity project was centred on providing a Web-based distance learning system, or 'Televersity', for a number of courses accredited by the Open College Network (OCN). *Solstra* provides the core administrative functions for the delivery of these courses. However, it is deficient in the critical areas of supporting groupwork and managing the evidence portfolio for each student. To meet this requirement the RISE components of the Meeting-Place (audio-conferencing), Workbooks (shared and private online folders for depositing evidence) and the People-Pages (information about the participants in each course) were integrated with *Solstra*.

The other key development is the ongoing work to extend the capability of the Generic Learning Platform (GLP). In this volume Fowler and Mayes describe the development of the theory of learning relationships as a guide for the design of new learning technology. In terms of the GLP we are now considering how this theory can be translated into the design of online learning systems, particularly focusing on moving beyond simple groupware to support the complex learning relationships within communities. We are now in the process of building our first prototype system based on this theory. We are following the same approach that was used in RISE, in that the core of the system is a data-model that adequately captures the rich set of objects, attributes and relationships that exist between stakeholders of the learning system. In addition a set of components is being constructed that will deliver parts of the functionality required. The key concept is the notion of the community, or 'group', and the relationships between participants in each group. At the time of writing we have completed the design of the data-model and the main system components and we are currently building the first prototype.

Contact author

Michael Gardner is a Technical Group Leader in the Internet and Multimedia Applications (IMA) unit at BT Adastral Park. In the past he has used expert systems and AI technologies to support customer service agents, and carried out research into object-orientated developments tools and methodologies. His work is now primarily focused on the use of the Internet, with a particular emphasis on distance learning and collaborative applications. He currently leads an 'Online Futures' research team.
[*michael.r.gardner@bt.com*]

Peripatetic electronic teachers in higher education

David Squires
School of Education, King's College London

Initially published in 1999

This paper explores the idea of information and communications technology providing a medium enabling higher education teachers to act as freelance agents. The notion of a 'Peripatetic Electronic Teacher' (PET) is introduced to encapsulate this idea. PETs would exist as multiple telepresences (pedagogical, professional, managerial and commercial) in PET-worlds; global networked environments which support advanced multimedia features. The central defining rationale of a pedagogical presence is described in detail and some implications for the adoption of the PET-world paradigm are discussed. The ideas described in this paper were developed by the author during a recently completed Short-Term British Telecom Research Fellowship, based at the BT Adastral Park.

Introduction

Most people would agree that the role of teachers in higher education will be changed by the advent of modern information and communication technology (ICT). At the very least open and distant learning (ODL) will become an established feature of educational provision rather than a specialist aspect. This is already evident in the increasing use of ODL in combination with conventional forms of course delivery by universities. More radically it is claimed that there will be a fundamental change in the nature of educational institutions, leading to the notion of virtual university, e.g. the Western Governors University (*http://www.wgu.ed*). These claims have in common the notion of utilizing the distributed nature of ICT to provide partnerships between teachers and learners that are not defined by either spatial or temporal constraints.

While the concept of a virtual university is radical, there is an implicit assumption that educators will still be primarily affiliated to one institution. An even more radical claim is that ICT will break exclusive links between educators and a single institution. Just as future learners may be seen as clients contracting to receive educational provision from a range of providers, educators may be seen as independent workers offering their services to learners on demand. These educators will not be confined to the classroom, rather they will be electronic workers providing a virtual presence in public spaces, the workplace and the home. A new type of *peripatetic* electronic teacher (PET) will emerge. Aristotle, the first peripatetic teacher, used to walk around the Lyceum as he taught; PETs will surf around the Net as they teach.

These changes were foreseen in principle long before the advent of the Web. In 1970 Ivan Illich (1971) introduced the idea of a 'learning web' providing global access to learning resources and enabling free communication between learners and teachers. A combination of the idea of a learning web and the possibilities afforded by ICT leads to the notion of a peripatetic electronic teacher. PETs will act as information brokers, providing routes for learners to search networked information systems. They will provide virtual asynchronous tutorials for small groups and individuals, act as the teacher in a virtual classroom environment, manage synchronous virtual seminars, and moderate discussion lists/bulletin boards. On occasion they will provide face-to-face teaching sessions, which will be advertised on the Web. Instead of using 'off-the-shelf' curriculum materials PETs will use multimedia authoring systems to design and publish materials, either for general circulation or as bespoke products developed in response to client demand.

Initial development work on a design specification and prototype for a PET software environment has been completed through a BT Short-Term Research Fellowship based at the BT Adastral Park. This paper describes this design specification.

A PET design rationale

Squires (1999) describes the original design rationale for a PET as an 'Assistant' providing a seamless interface to a diverse range of design, navigation and communication tools:

- A peripatetic network-based teacher will need to maintain an overt presence on the Web. At a practical level this involves being able to handle diverse requests from clients, advertise services and communicate with colleagues.

- Some teachers may wish to combine a Web-based virtual presence with contract teaching commitments in diverse locations. These teachers will need to be contactable in a variety of locations.

- Brokering information will require access to state of the art network navigation systems, and possibly the development of navigation tools customized to the needs of education. Navigation tools with different approaches will need to be presented in a coherent and synergistic way.

- Designing learning materials will require access to a range of complementary authoring environments.

- Digital publishing will need support – copyright will need to be protected, payments will need to be made and links with software publishing agencies will need to be maintained.

This is a utilitarian and functional view which originates from a view of the Internet/Web as primarily an information and distribution mechanism. However, the 'superhighway' metaphor is immature – it is being replaced by the notion of global networks as social places. Schlager *et al* (in press) cite McFarlane (1996) in support of this emerging notion; they regard a networked environment as a gathering place, or agora, that 'brings people together, encourages participation, and supports creativity, a place that is always growing, adapting, and changing in response to new ideas and initiatives'. This leads to a rather different perception of the design of a PET Assistant. It is not a well-ordered and comprehensive toolbox that is needed, but rather a flexible virtual space/environment which allows a PET to have an effective and functional telepresence on the Web. In this sense it is a question of designing a PET-World, inhabited by PETs and learners, rather than a PET Assistant.

With this perspective in mind a design specification for a PET-World has been developed based on the idea of four presence domains:

- *Pedagogical presence* where the PET appears as a teacher, playing roles such as instructor, coach, mentor, tutor, scaffolder and expert.

- *Professional presence* where the PET appears as a member of the teaching profession, playing roles such as colleague, committee member, trainer and trainee.

- *Commercial presence* where the PET appears as a freelance worker available for hire, playing roles such as consultant, personal tutor and publisher.

- *Managerial presence* where the PET appears as an administrator, scheduling teaching commitments, validating learners attendance/achievements and managing course enrolment.

All domains are important in realizing a holistic perception of the role and scope of a PET-World, and initial design specifications exist for each domain. As PET-Worlds are developed other presences may be required. However, the pedagogical domain provides a central defining rationale and the specification for this domain has been developed to the greatest extent. The remainder of this paper will describe this specification and reflect on the implications of implementing the design.

Pedagogical telepresence

Teachers communicate with learners through two fundamental channels: exposition and discourse. Pedagogical presence is defined by the character of exposition to learners and the nature of discourse with learners. From a constructivist perspective exposition is not simply a question of presenting ideas, facts and so on for consumption, i.e. pedagogical presence is not simply presentational. Rather it is a question of exposing learners to learning situations which encourage them to develop ideas and concepts in ways which match their idiosyncratic needs and demands. Thus exposition will be a mix of presenting ideas and concepts for consideration, setting problems to be solved and organizing projects. An essential aspect of good teaching is feedback and advice based on effective assessment. Discourse provides a way of giving this feedback, with the teacher assuming roles such as guide, mentor, adviser, moderator, critic, leader and instigator in discussions – both between teacher and learner(s) and between learners (peer group collaboration). In this sense discourse provides opportunities for the teacher to scaffold learners' activities.

Exposition and discourse will remain as defining communication channels for PETs, but the nature and scope of these channels will change, e.g. asynchronous conferencing

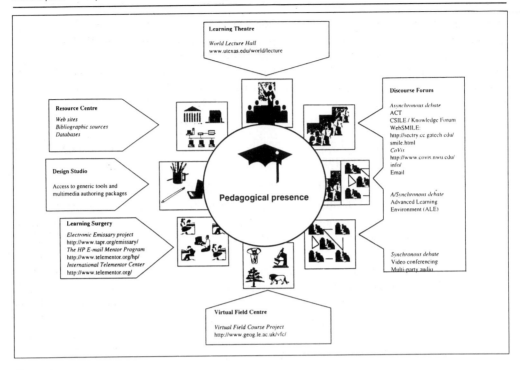

Figure 1

systems are introducing new aspects of discourse between teacher and learner. A consideration of exposition and discourse in a PET-World context suggests a number of 'learning rooms' within the Pedagogical Presence Domain (see Figure 1):

- a *Learning Theatre* in which exposition can take place in an interactive distributed multimedia environment;

- a *Discourse Forum* based on access to a range of synchronous, asynchronous and hybrid a/synchronous systems providing opportunities for the PET to scaffold learners' activities;

- a *Virtual Field Centre* in which the PET acts as guide and designer of thematically constructed explorations of the Web;

- a *Learning Surgery* in which a PET can assume the roles of tutor and mentor to respond to requests for specific help and advice form individuals and small groups;

- a *Design Studio* providing a PET with access to a range of design tools to develop curriculum materials;

- a *Resource Centre* where a PET is a guide to the selection and retrieval of information.

Learning theatre

Providing access to diverse, extensive and high-quality materials to support networked lecture presentations is an immediately obvious rationale. Didactic presentations of knowledge can be augmented by use of slides, video, animations and simulations – very much as in an Open University television programme. Basically this rationale rests on the idea of a (multimedia) replication of conventional lecture presentations.

This is a limited view which does not take account of a constructivist learner-centred focus. Norman (1998), in describing the use of multimedia networked classrooms at the University of Maryland, observes that the trend has been to 'evolve from presentations to explorations, from passive learning to active engagement and from "sage on the stage" to "guide on the side"' (p. 42). Norman cites Shneiderman *et al* (1995) as emphasizing this trend in stating that the electronic classrooms at the university were originally called 'teaching theatres', but with experience were renamed 'learning theatres'. This term is borrowed to describe an area within the Pedagogical Presence Domain.

A learning theatre will combine the possibilities of hi-fidelity presentation with synchronous interaction, leading to the notion of *participative exposition*. Learning theatres will be examples of shared spaces as described by Bradley *et al* (1996). The inhabitants of these spaces will be the teacher(s) and the learners. Interaction will typically be orchestrated by the teacher, but there will be possibilities for learner-learner interaction as well. Emerging virtual reality technologies provide exciting possibilities in this context, e.g. systems such as *The Mirror* have a three-dimensional shared space (Walker, 1997) which combines multiple avatar presence with synchronous text-based discourse. Future worlds may include multi-party audio. It is not hard to imagine a learning theatre populated by avatars (teacher(s) and learners) with real time audio conferencing.

Identity
A conventional classroom provides teachers with a social space in which they can project their personality and exercise their role as a teacher, i.e. a forum for maintaining a pedagogical presence. Learning theatres will need to afford PETs similar possibilities. Personal appearance may offer one way of providing a *personality affordance*. For example, 'The Mirror included a choice of four avatars with clothing which could be coloured from an extensive palette' (Walker, 1997, p.32). A similar approach could be adopted in learning theatres, with PETs offered the facilities to design their own appearance and, if multi-party audio is used, their own voice characteristics. If synchronous text based discourse is employed, there may be some merit in presenting text contributions as characteristic handwriting rather than as typed script. Teachers and learners could choose a handwriting style from a style library. As they typed in text it could automatically be converted into script presented in this style. Another possibility is the availability of a voice to text system for use by the PET.

Learning theatre management
Well run classrooms are characterized by diverse practice: whole group presentation, small group work, individual work and circus activities. Learning theatres will need to offer PETs similar *management affordance*. The use of 'auras' (grouping algorithms used to produce scaleable worlds) could be relevant in this context. These algorithms have been developed to make shared spaces populated by large numbers of avatars technically feasible and practically manageable by users. Users are placed in need-to-know groups, enabling the system to control the amount of information sent to any user. The use of auras could also open up the possibility of *massive participatory exposition*, i.e. lessons in which large

numbers of students could participate. These could be celebrity lessons given by 'excellent PETs'.

Motivation

Trials of *The Mirror* indicate 'the importance of special events and the pulling power of celebrities in maintaining the community' (Walker, 1997, p. 37). The notion of massive participatory exposition is relevant here. This raises issues of PETs working in teams. PETs would spend most of their time working with 'their' class, but would also act as an associate PET for special events. Of course, there are possibilities for team teaching as well.

Authority

What are the new forms of misbehaviour and ill-discipline that will arise in a virtual environment? Vandalism may take the form of deliberate malicious annotations to materials presented in open collaborative environments – multimedia graffiti. It may take the form of deleting and/or changing other peoples' work. Subtle changes may be the most insidious. Another form of ill-discipline may be unacceptable behaviour in open environments (e.g. MUDs) or the use of unacceptable language (e.g. in discussion groups). Will 'flaming' take on a new meaning? Hacking into confidential systems (e.g. learner records) may be a problem. Virtual discipline will raise a number of software development issues, e.g. providing controlled access to the Web, creating 'protected' user groups, and allowing censorship through blocking and regulatory systems (Martin, 1997).

Discourse forum

In conventional educational settings discourse is synchronous: classroom discussion, seminar groups, conversations between teachers and learners. While new technology opens up possibilities for replicating synchronous discourse (video conferencing, multi-party audio, telephone conferencing and text-based chat lines), perhaps the most interesting development is the emergence of asynchronous discourse systems. The fundamental claim here is that asynchronous discourse encourages reflection. Of course, the two forms can be combined to give a/synchronous discourse systems.

The notion of a discourse forum does not have a conventional equivalent, which attempts to provide a space defined in terms of a mixture of synchronous and asynchronous debate and exchange of information. A number of design issues have emerged in the design and development of systems such as those which would feature in a discourse forum: annotated discourse, configured discourse, and multimedia discourse.

Annotated discourse

It is commonly argued that asynchronous discussion fosters reflection leading to more thoughtful and considered responses. This begs the issue of how learners can adequately represent their reflective thinking. One common approach is to provide labelling facilities, e.g. as in the Asynchronous Conferencing Tool (ACT) described by Sloffer *et al* (1999). In ACT messages can be labelled as hypothesis, important point, evidence, and learning issue. In some systems there are reserved labels for teacher use only, e.g. a problem-setting label or a summary label.

Configured discourse

Asynchronous discussions can also be recorded allowing participants to look at the history of a discussion. This in turn leads to attempts to represent discourse history. A variety of techniques exist. Discourse history can be represented in terms of argument threads, tasks or simply as a temporal record. Each of these approaches has its advantages and dis-advantages.

Multimedia discourse

A number of systems, e.g. the Collaborative and Multimedia Interactive Learning Environment (Hübscher *et al*, 1997), allow multimedia annotations. It is likely that the nature of discourse labels will change to encompass multimedia objects in addition to text: voice notes, video, graphics, pointers to URLs, etc.

Virtual field centre

Project-based activities involving information and data-gathering, analysis, synthesis and report are commonly advocated as a means of providing authentic learning experiences. Field-trip-based activities are a common manifestation of a project-based approach – visits to natural habitats to explore ecological issues, visits to museums, art galleries, visits to foreign countries, etc. In a Web context it is possible to think of virtual field trips: guided explorations of natural habitats which would include simulations, visits to resource centre Web sites, and so on; art trips involving virtual visits to thematically selected exhibitions; historical trips involving visits to museums; and cultural trips to other countries with an emphasis on communication in the local language. The JISC-funded Virtual Field Course Project (*http://www.geog.le.ac.uk/vfc*), featuring collaboration between the University of Leicester, Birkbeck College, Oxford Brookes University and City University, provides an example of a development in this area.

In a virtual field trip the PET would have two roles: (i) designing and organizing the trip and (ii) acting as guide and navigator during the trip.

Navigation design

The design of the trip will involve the identification of relevant sites and proposing a route(s) for learners. Maps, signposts and itineraries all suggest themselves as navigational devices. However, as Platt and Willard (1998) point out, these techniques are not sensitive to temporal changes:

> There are many techniques used in existing virtual environments that aim to reduce this risk of 'getting lost', such as signposts, maps, landmarks and viewpoints. However, these are generally set when the environment is originally designed and based on the designer's view of what is an interesting part of the world. Crossley *et al* (1997) predict that 3D virtual environments will become more organic, being built and updated dynamically, with greater personalisation of the worlds and greater interactivity with objects. Therefore, designer-driven navigation aids may not be able to reflect such developments in the environment brought about by user interaction or changing information.

This is a serious design issue for virtual field trips. In these trips there should be possibilities for learners to explore by adapting/modifying the route(s) so as to foster serendipitous learning: they should not be constrained by the designer's original intentions. These adaptations in themselves provide a rich set of educational materials, implying that the virtual field trips should be mutable (Anderson and McGrath, 1997) to reflect changes in the range of suitable locations over time and the experiences of previous learners.

Platt and Willard (1998) have proposed three techniques that 'would be able to evolve along with the space and the community':

- The idea for *virtual cairns* is derived from the piles of stones that hill walkers place to mark routes and indicate points of interest. Platt and Willard describe virtual cairns as viewpoints that can be set by users rather than the designer. With a viewpoint there is an associated object or collection of objects, allowing users 'to see which positions or routes are important to other users, so the whole community contributes to the understanding of the space'. If the movement of users is monitored virtual cairns could be placed at frequently visited locations. It would be possible to link virtual cairns to other resources – video, text, etc. In a virtual field trip the temporal and spatial sensitivity of virtual cairns, the implicit collaboration between learners in adding and changing cairns, and the possibilities afforded by multimedia resource links to cairns make them a very attractive design technique.

- *Trails* correspond to the routes most frequently taken by users. These can be generated by automatically tracking the routes of users and synthesizing the most popular routes. They can be represented graphically in a virtual space. In a virtual field trip some default routes could be included to act as an introduction for learners.

- *Intelligent tours* are based on the idea of storing how people use a space and using this information to propose the most interesting route. The Information Garden (Crossley *et al*, 1997) provides an example of an intelligent tour. Platt and Willard (1998) describe the simplest case of an intelligent tour as when 'the system might only be aware of the location of objects or information in the space, and move the user via these locations'. They propose that tracking other people in the environment and being aware of popular routes and locations, recent changes in objects and information, and the specific interests of users would make the system more useful. This would allow the creation of 'a tailor-made route for any particular user, ensuring that they discover any relevant parts of the environment (information or people) whilst moving quickly and effectively through the space' (Platt and Willard, 1998). Intelligent tours could open up the possibility of matching virtual field trips to the needs of individual learners.

Learning surgery

A significant component of students' total learning experience is access to dedicated teaching support on a personal basis. In this context learners have a chance to express their own learning difficulties and get help geared to their own needs. Personal teaching in a conventional context is typically provided by a tutor working with individuals or small groups. The notion of mentoring, with subject experts offering individual learners advice and information, is also relevant in this context. There is a clear role for PETs to act as tutors, utilizing networked conferencing systems and email to provide synchronous and asynchronous support at a personal level. Tutoring by PETs leads to the idea of a Learning Surgery where individual learners can visit a PET for personal assistance. Conferencing systems also open up the possibility of putting learners in contact with expert mentors from universities, industry and commerce.

Individual and small group tutorials

Synchronous communication, either between a PET and an individual learner or a small group of learners, offers possibilities of replicating many of the aspects of conventional personal and small group tutorials. Collaborative artefacts, such as interactive whiteboards and shared real-time documents and files, will be very important in the short- and medium-term future. On a longer time-scale shared virtual spaces, with participants represented by avatars, may provide the design rationale for the development of a Learning Surgery in a PET-World.

Asynchronous conferencing systems and email could also be used to conduct personal and small group tutorials. The advantage here might be an emphasis on reflective thinking. These systems could also provide an 'out of hours' emergency service. For example, learners having trouble with assignments could post a request for an urgent synchronous appointment with a PET to discuss the problem. Perhaps a bulletin board system could be used which would allow learners to page a PET by using a special code.

Telementoring

An example of how access to networked conferencing tools can facilitate mentoring is given by work at Georgia Institute of Technology. In an architecture class, students were asked to use *CoWeb* (a co-operative Web-based environment) to create two pages for each project they completed: (i) a project description with scanned images of their drawings (pin-ups) and (ii) a statement of research questions to which they needed answers in order to complete their designs. Guzdial (1998) describes how expert architects acted as mentors by using *CoWeb*:

> Expert architects were invited to tour the students' pin-ups and comment on the projects. For each expert architect, a 'tour page' was set up with the architect's name on it. The architect was invited to visit each of the pin-up pages listed on his or her tour, and comment on the pin-ups either directly on the student's page or on the tour page. This activity was perceived as being more successful in supporting the students' learning. The experts wrote a surprising amount of commentary. They sometimes left comments on students' pin-up pages with particular advice, and sometimes they wrote on the tour page with general advice that the expert felt that the group needed. Students took the reviews quite seriously and worked to respond to the experts' comments. (Guzdial, 1998)

Design studio

The open malleable nature of distributed multimedia provides opportunities for PETs to act as the designers of multimedia educational materials. In this capacity they will need to have access to a range of multimedia authoring tools. Within the Design Studio it is possible to envisage a number of ways of providing access:

- links to locally stored generic tools (e.g. DTP packages, modelling systems, spreadsheets) and authoring packages (e.g. systems such as *Authorware Professional, Director*, etc.);

- links to Web sites offering shareware design and development tools;

- links to the Web sites of commercial vendors of design/development tools;

- links to collaborative design and development environments.

Resource centre

Learners are going to be confronted with ever increasing amounts of information, and an expectation that their work should be characterized by information-rich exploration and expression. They will need help in selecting information and resources. There is a role here for the PET as an information consultant – someone who will point learners in the direction of useful and relevant resources. In this sense PETs will act like librarians.

At a simple level this support could just take the form of a catalogue of useful resources (Web sites). However, there is a need to customize and personalize support. PETs could give advice on where to search for relevant information and on appropriate search engines to use. Perhaps they could engage in collaborative searching, using shared applications. Intelligent agents may be very useful in the future. Perhaps the possibility of a mixture of PET and intelligent agent support would be a good idea.

One design possibility centres on the concept of mirror virtual worlds which 'provide a second-person experience in which the viewer stands outside the imaginary world, but communicates with characters or objects inside it' (McLellan, 1996, p. 460). A digitized video image of the user is generated and superimposed on the virtual world with its objects and characters, allowing an interaction between the user and the inhabitants of the virtual world. Lewis and Cosier (1997) refer to a mirror system for virtual conferences which mixes real video, avatars, and a pictorial representation of an intelligent agent. In such an environment a PET could project his or her presence into a virtual resource centre visited by avatars representing learners in need of advice about resource and/or information location. A learner's avatar and a PET could communicate to establish the information needed and how best to obtain it. On the basis of these decisions an intelligent agent could be configured to effect the search and deliver the results.

PET-Worlds in higher education

Is the concept of a PET-World relevant in higher education systems which are being metamorphosed by radically changing attitudes, resource levels and expectations? How would PETs relate to the rapid expansion of higher education, with the implied need to cater for significantly increased student numbers from varied cultural and educational backgrounds? Would PETs help to meet demands for improvements in the quality of university teaching or would they compromise attempts to improve quality? Would there be possibilities for improving cost-effectiveness and better use of resources?

There is clearly a simplistic argument that the adoption of a paradigm based on the extensive use of distributed global networked systems will inevitably lead to improved access, making higher education more available to more people in a variety of locations unbounded by temporal constraints. However, this is a superficial argument which makes a number of assumptions. First, there is an assumption that access to the technical infrastructure supporting PET-Worlds will be widespread, if not universal. While information and communications technologies are becoming pervasive, it is probable that in the foreseeable future some potential students will not have access to appropriate computing facilities. In addition, will all students be able to afford PET fees? Thus the potential to improve access implies significant equity issues.

What mechanisms and funding issues need to be considered to ensure equitable access? In some ways the establishment of the University for Industry and the National Grid for Learning presage the type of developments that would be necessary to encourage the development of PET-Worlds. Other funding models may introduce equity problems. For example, if PETs become common it is reasonable to assume that there would be an increase in telecommunications traffic resulting in increased revenue. Increased use of infrastructure will result in more demand for hardware and more demand for better connectivity (extent and speed). In this context infrastructure providers may assume a role as a third party broker of educational services, with implications for the price of accessing PET services.

As access becomes more universal and varied in character, conventional institutions may be placed under threat, e.g. through the widespread development of agencies which broker the services of self-employed educationalists. This is very much the model adopted by the widely quoted University of Phoenix. This could lead to a gradual decline in standards. What quality-assurance procedures would need to be established to counter any such decline?

The university curriculum may be compromised, with only those subjects offered by PETs which fit easily into the PET-World framework. The prospectuses of existing virtual universities indicate that this could be a problem, with a very heavy emphasis on vocational subjects such as business studies, nursing and education, to the exclusion of specialized academic subjects such as classics and nuclear physics. Would the widespread adoption of the PET paradigm lead to an expansion in vocational education at the expense of academic scholarship?

The establishment of PET-Worlds implies a large audience for educational materials, which in turn implies a cost-effective approach to resource provision. In particular the arguments rehearsed by Mayes and Fowler (1999) for the reuse of students' materials become very attractive. However, the initial cost of the development of high-quality technology-based learning materials should not be underestimated. These initial costs will need to feature in any financial models for the development of PET-Worlds.

Clearly this discussion is limited and speculative, based as it is on a design specification and limited prototype development. Nevertheless, it points up some important potential advantages and disadvantages of embarking on the developments of PET-Worlds. While there are potential difficulties and pitfalls the PET paradigm could offer much in helping universities to cope with a radically changing higher education context.

References

Anderson, B. and McGrath, A. (1997), 'Strategies for mutability in virtual environments', *Proceedings of Virtual Environments on Internet, WWW, and Networks, National Museum of Photography, Film and Television*, Bradford,
http://www.visual.bt.co.uk/users/abc/andersb2/public/papers/SfM.doc.

Bradley, L., Walker, G. and McGrath, A. (July 1996), 'Shared spaces', *British Telecommunications Engineering Journal*, 15, *http://vb.labs.bt.com/msss/IBTE_SS/.*

Crossley, M., Davies, N. J., Taylor-Hendry, R. J. and McGrath, A. J. (1997), 'Three-dimensional Internet developments', *British Telecom Technology Journal*, 15 (2), 179–93.

Mayes, J. T. and Fowler, C. J. (1999), 'Learning technology and usability: a framework for understanding courseware', *Interacting with Computers,* 11 (5), 485–97.

Guzdial, M. (1998), 'Collaborative websites to support an Open Authoring Community on the Web', *http://guzdial.cc.gatech.edu/papers/CoWeb/*.

Hübscher, R., Puntambekar, S. and Guzdial, M. (1997), 'A scaffolded learning environment supporting learning and design activities', *Proceedings of the American Educational Research Association Meeting 1997*,
http://guzdial.cc.gatech.edu/papers/aera97/scaffolding.html.

Illich, I. D. (1971), *Deschooling Society*, Harmondsworth: Penguin Books.

Lewis, A. V. and Cosier, G. (1997), 'Whither video? – pictorial culture and telepresence',
British Telecom Technology Journal, 15 (4), 64–85.

Martin, D. (1997), 'Putting teachers and parents in control: Internet content labelling and blocking technologies', *Proceedings of the National Educational Computing Conference*, Seattle, Washington, USA, 281–6.

McFarlane, M. C. (1996), 'Humanizing the superhighway', *Technology and Society*, 14 (4), 11–18.

Norman, K. (1998), 'Collaborative interactions in support of learning: models, metaphors and management', in Hazemi, R., Hailes, S. and Wilbur, S. (eds.), *The Digital University: Reinventing the Academy*, London: Springer, 39–53.

Platt, P. and Willard, A. (1998), 'The ramblers guide to virtual environments', *IEE Colloquium: The 3D Interface for the Information Worker* (May 1998),
http://www.hfnet.bt.co.uk/projects/IEE_paper/IEE_paper_v1.htm.

Schlager, M., Fusco, J. and Schank. P. (in press), 'Conceptual cornerstones for an on-line community of education professionals', *IEEE Technology and Society*,
http://www.tappedin.org/info/ieee.html.

Shneiderman, B., Alavi, M., Norman, K. and Borkowski, E. (1995), 'Windows of opportunity in electronic classrooms', *Communications of the ACM*, 38, 19–24.

Sloffer, S., Duebur, B. and Duffy, T. (1999), 'Using asynchronous conferencing to promote critical thinking: a case study of three implementations in higher education', *Proceedings of the Hawaii International Conference on Systems Science*, (CD–ROM), Piscataway, NJ: IEEE Computer Society Press.

Squires, D. (1999), 'A teacher's PET for the millennium', in Tuomi-Kyro, E., Salonen, M., Saarinen, P. and Sinko, M. (eds.), *Communications and Networking in Education: Learning in a Networked Society Conference Proceedings*, Aulanko-Hameenlinna, Finland, 328–33.

Walker, G. (April 1997), 'The mirror – reflections on inhabited TV', *British Telecommunications Engineering Journal*, 16, 29–38,
http://vb.labs.bt.com/msss/IBTE_Mirror/intro.htm.

Update – Peripatetic electronic teachers in higher education

The idea of peripatetic electronic teachers (PETs) is meant to be provocative. The aim of outlining a design specification was to stimulate innovative thinking about the changing roles of teachers as global networks become ubiquitous. This specification does not take account of the many problems that would be encountered in the design and implementation of PET-worlds. For example, in a recent interview for the *Connected* electronic magazine (*http://www.connected.org/learn/david.html*) I was asked 'Amongst the functions of existing higher-education institutions, there is a certain label of credibility given to those working for them. In a system in which many teachers are independent agents, how do you propose to guarantee the quality of teaching?' For what it is worth, my answer was:

> You are quite right, at the moment most teachers are legitimized by their host institution; remove the teachers from the institution and you remove their formal status. Perhaps third-part agencies could be established to licence PETs. These could include a range of organizations, including infrastructure providers such as Telecom companies. Perhaps licensed PETs could be allowed to add credit/comments to the digital learning portfolios that we will all carry around with us in the future? Another, more radical approach would be to adopt a market perspective – good PETs would thrive, poor PETs would not be employed and cease to operate.

This is radical stuff that may not be to everyone's taste. However, there are signs that PET-like ideas are emerging. In *The Times* (4 November, 1999) it was reported that 'Pupils will be able to search the Internet for lessons from the best teachers under a £3 million "virtual classroom" approved by the Government yesterday' (p. 2). In a higher-education context the *Times Higher Education Supplement* (7 January 2000) reported under the headline 'How to hold a tutorial in cyberspace' the development by the Open University of the *Lyceum*. Using *Lyceum* tutorials can be given at a distance using synchronized voice and visual conference software including a shared whiteboard, a concept mapper and a screen grabber. While the advent of PETs may be some time off, there are signs that some the aspects in the PET design specification are becoming evident in learning technology developments.

Contact author

David Squires is a Professor of Educational Computing, School of Education, King's College London. His research and development interests are the design and evaluation of educational software, teaching in networked environments and the use of ICT in academic research. [*david.squires@kcl.ac.uk*]

Index of articles

(with original publication details)